Building-Integrated Photovoltaics

Building-integrated photovoltaics (BIPV) is an innovative technology offering a variety of building envelope solutions, materials, and colours for virtually any building surface. These BIPV products generate on-site renewable electricity, turning buildings from energy consumers into producers. BIPV is expected to play an indispensable role in the transition towards decarbonisation and energy resilience of cities, effectively reducing energy consumption and greenhouse gas emissions. Lack of knowledge and guidance on designing BIPV systems has hindered this technology's widespread adoption and creative applications. As a remedy, this guidebook presents best practices and decision-making processes for efficient and resilient architecture. Featuring more than 50 annotated reference drawings – roofs, solar shadings, rainscreen façades, curtain walls, and double-skin façades – and 24 international BIPV case studies, the guidebook provides building professionals with the technical knowledge and inspiration to implement BIPV technology in the built environment.

Nuria Martín Chivelet, Ph.D. in physics, has developed her activity in photovoltaics at the Spanish research centre CIEMAT (www.ciemat.es). She has been involved in more than 30 national and European research projects, with scientific and technical production on characterisation and modeling of PV and BIPV modules and systems (https://orcid.org/0000-0003-4224-6618). In parallel, she has been a member of several national and international experts' committees on PV and BIPV, such as the standardisation IEC-TC82, CENELEC-CLC/TC 82, and several tasks within the International Energy Agency PV Power Systems Program (IEA-PVPS) related to BIPV, co-leading Subtask "BIPV Guideliness".

Costa (Konstantinos) Kapsis, Ph.D., is an assistant professor in the Department of Civil and Environmental Engineering at the University of Waterloo. He is a building scientist and an architectural engineering educator working in the field of high-performance building design, with a focus on electrification of the built environment, advanced building envelope technologies, digital twins, and daylight. Costa has been involved in the design of more than 30 net-zero buildings. He is the co-author of several peer-reviewed publications, book chapters, and international reports and standards.

Francesco Frontini, Professor, graduated in construction engineering and architecture, and collaborated with engineering and architecture firms as a project manager from 2007 to 2010. In 2009 he obtained a Ph.D., developing a new multifunctional BIPV façade for solar control. He worked as a post-doc in the solar façades group at Fraunhofer ISE (Germany), one of the world's largest research institutes. Since 2011, he has been working at the University of Applied Sciences and Arts of Southern Switzerland (SUPSI). He is Professor of Innovative Building Technology and Sustainable Design and is responsible for the sustainability area of the Department of Architecture and Construction. His research activity has always been supported by experimental work on the design of real buildings and solar envelopes. He is a member of the standardisation bodies SIA, CENELEC, and ISO. Since 2021, he has been appointed by the Swiss Federal Office of Energy as Co-Manager of the IEA PVPS Task 15.

Building-Integrated Photovoltaics

A Technical Guidebook

Edited by Nuria Martín Chivelet, Costa Kapsis
and Francesco Frontini

Routledge
Taylor & Francis Group
NEW YORK AND LONDON

First published 2025
by Routledge
605 Third Avenue, New York, NY 10158

and by Routledge
4 Park Square, Milton Park, Abingdon, Oxon, OX14 4RN

Routledge is an imprint of the Taylor & Francis Group, an informa business

ISBN: 978-1-032-55789-2 (hbk)
ISBN: 978-1-032-51705-6 (pbk)
ISBN: 978-1-003-43224-1 (ebk)

DOI: 10.1201/9781003432241

Typeset in ITC Officina Sans
by Apex CoVantage, LLC

Contents

Foreword vii
Preface viii
Acknowledgments x
List of contributors xi

CHAPTER 1 **Introduction** 1

CHAPTER 2 **BIPV performance requirements** 3
2.1 Introduction 3
2.2 Electricity generation 3
2.3 Thermal performance and solar gain 7
2.4 Daylighting and visual comfort 9
2.5 Acoustic insulation 11
2.6 Durability, reliability, and safety 12
2.7 Aesthetics and architecture 16
References 20

CHAPTER 3 **BIPV products** 24
3.1 Introduction 24
3.2 Components and assembly of a BIPV module 24
3.3 BIPV envelope solutions 32
References 40

CHAPTER 4 **A decision-making process for BIPV design** 42
4.1 Introduction 42
4.2 Step 0: Why BIPV? 43
4.3 Step 1: Assess building site 43
4.4 Step 2: Perform a solar access study 44
4.5 Step 3: Determine annual BIPV energy generation target 45
4.6 Step 4: Estimate BIPV installed capacity and energy yield 47
4.7 Step 5: Develop the BIPV design 48
4.8 Step 6: Assess sustainability, circularity, and life cycle cost 50
4.9 Business models 53
References 54

CHAPTER 5 **Design of BIPV envelope and case studies** 56
5.1 Introduction 56
5.2 BIPV roof systems 58
5.3 BIPV façade systems 106

5.4 Externally integrated systems 150
5.5 Infrastructure 160
References 166

CHAPTER 6 **Operation and maintenance of BIPV systems** 167
6.1 Introduction 167
6.2 Main components of an O&M plan 167
6.3 Safety aspects 168
6.4 BIPV system operation 169
6.5 BIPV system maintenance 171
References 172

Annex I: Symbols, acronyms, and abbreviations 173
Annex II: List of referenced standards 176
Index 177

Foreword

The world of solar energy is currently experiencing a decisive turning point at the global level by becoming one of the key players in the energy transition and the fight against the effects of climate change. At the end of 2023, it is estimated that nearly 1.5 terawatts (TW) will be installed worldwide, which now represents a significant share of the global energy mix. At the same time, solar energy constitutes a significant economic force, with more than $US 200 billion in turnover worldwide. The building sector, which constitutes half of the world's energy consumption, is of course not spared from this "solar wave," because this technology has multiple advantages for integrating energy production within the structure of a house, a hangar, offices, or even collective residential buildings. Any surface of a building with sun exposure will have to be questioned on the relevance for the installation of solar panels in the years to come, as the need is so important in the effort to decarbonise the sector.

This is where the wonderful idea of architectural integration of solar energy, otherwise known as Building-Integrated Photovoltaics (BIPV), comes into play.

This specific edition of the *IEA PVPS Task 15 BIPV Guidebook* has been prepared under the leadership of three specialists of this technology – Nuria Martín Chivelet, Costa Kapsis and Francesco Frontini – together with a team of worldwide experts comprising nearly 20 authors from 10 countries working within the framework of the International Energy Agency's Photovoltaic Power Systems Programme (IEA PVPS) Technology Collaboration Programme Task 15 on "Enabling Framework for the Development of BIPV." Building on the previous and long experience from Task 15 on this topic of BIPV, this handbook is a prominent example of technology and scientific collaboration across the BIPV field. It is aimed at covering three decisive objectives: (1) consolidate existing BIPV industry knowledge; (2) support the implementation of best BIPV practices (for new and retrofit buildings); and (3) drive the decision-making process that could lead to an effective BIPV design and a robust BIPV installation.

The heart of this guidebook, made of numerous well-documented case studies on BIPV, will surely deliver a lot of ideas for architects, building owners, promoters, and building engineering specialists so as to implement solar panels and solar energy-conciliating architecture and clean energy production. This is where this guidebook meets one of the primary IEA PVPS program objectives: enhance the international collaborative efforts that facilitate the role of photovoltaic solar energy as a cornerstone in the transition to sustainable energy systems.

The IEA PVPS Technology Collaboration Programme is pleased to publish the edition of this handbook together with this high-level group of specialised experts. Most importantly, I would like to acknowledge the leadership of Francesco Frontini, IEA PVPS Task 15 Manager, and the IEA PVPS Task 15 experts.

The IEA Photovoltaic Power Systems Programme (PVPS) is one of the Technology Collaboration Programmes (TCP) established within the IEA, and since its establishment in 1993, the PVPS participants have been conducting a variety of joint projects in the application of photovoltaic conversion of solar energy into electricity. As one of the few truly global networks in the field of PV, IEA PVPS can take a high-level, strategic view of the issues surrounding the continued development of PV technologies and markets, thus paving the way for appropriate government and industry activity. Among decisive topics such as grid integration, sustainability, and reliability, integration into the buildings is a key matter for PVPS, and Task 15 has been perfectly addressing this challenge for seven years now.

I hope this guidebook finds many interested readers and contributes to the further deployment of solar energy worldwide.

Daniel Mugnier – Chair,
IEA PVPS Technology Collaboration Programme
January 2024

Preface

In the continually evolving landscape of sustainable architecture and renewable energy integration, the realm of building-integrated photovoltaics (BIPV) stands as a beacon of innovation and promise.

BIPV, in its essence, embodies a transformative approach to architecture and energy conversion. It seamlessly integrates solar technology into the very fabric of buildings, converting them into active contributors to the renewable energy landscape. This synergy between architectural design and solar electricity generation not only addresses the growing demand for sustainable building practices but also presents a novel solution to the challenges posed by traditional energy sources. The challenges and opportunities inherent in BIPV are multi-faceted. On the one hand, BIPV offers a promising path to reduce our reliance on conventional energy sources (still based mainly on fossil fuels), fostering a more sustainable and ecologically friendly built environment. On the other hand, integrating solar technologies into architectural elements demands a nuanced understanding of both disciplines – architecture and photovoltaics. Balancing the aesthetic and functional aspects of BIPV installations while optimising their performance requires skilful interaction between technical expertise, design innovation, and a keen awareness of the unique challenges presented by each project.

This guidebook, a product of collaboration among experts from various regions of the globe, emerges as a comprehensive resource dedicated to explaining and streamlining the technicalities of holistic design and processing of BIPV systems for solar architecture. As the second publication of the International Energy Agency Photovoltaic Power Systems Programme (IEA-PVPS) Task 15 that presents real BIPV case studies, it signifies a significant leap forward in the understanding and application of BIPV in the built environment.

IEA-PVPS Task 15, titled "Enabling Frameworks for the Development and Implementation of BIPV," served as the foundational platform for the creation of this guidebook. The task, established to address the challenges and opportunities inherent in BIPV deployment, brings together international experts to share knowledge, best practices, and innovative solutions. The workplan of the task started in 2016 with the first phase and will continue for a third phase from 2024 until 2027. It is organised today in several activities covering the broad spectrum of BIPV: Technological Innovation System (TIS) Analysis for BIPV; cross-sectional analysis: learning from existing BIPV installations toward a multi-dimensional assessment approach; BIPV Guidelines; digitalisation for BIPV; and pre-normative international research on BIPV characterisation methods. The collaborative spirit of Task 15 has fuelled the insights and recommendations found on the guidebook pages, witnessing to the collective efforts of professionals dedicated to advancing the field. In a previous publication, available online on the home page of IEA-PVPS Task 15, we collected and shared the experience of people who integrated photovoltaics into the building envelope, and illustrated how such buildings can appear.

As architects and professionals in the building sector grapple with the imperative to integrate renewable energy solutions into their designs, subtly or boldly, this new book offers a wealth of insights, drawing from the collective wisdom of international contributors. The chapters within this guidebook are meticulously structured to cater to the specific needs and interests of the target audience. From establishing fundamental performance requirements (Chapter 2), through navigating the diverse landscape of BIPV products (Chapter 3), to delving into a systematic decision-making process for BIPV design (Chapter 4), the book unfolds as a roadmap for navigating the complexities inherent in BIPV implementation.

Chapter 5 immerses the reader in the design complexity of BIPV envelopes, offering a detailed review of case studies that exemplify successful applications. Finally, Chapter 6 sheds light on the often-overlooked aspect of operation and maintenance, ensuring that BIPV systems not only impress on their inception but also endure as reliable and efficient contributors to the sustainable built environment.

The primary objectives of this guidebook are clear: (1) to consolidate the existing knowledge within the BIPV industry; (2) to provide robust support for the implementation of best practices in both new constructions and retrofit projects; and, above all, (3) to empower decision-makers in the field with the tools necessary for effective BIPV design.

In a world where the sustainable evolution of our built environment is non-negotiable, this publication serves as an invaluable resource, offering not just theoretical insights but practical wisdom garnered from real-world case studies. As we collectively strive to usher in an era of energy-efficient, aesthetically pleasing and environmentally conscious buildings, the technical guidance within these pages stands as a beacon, directing architects and building professionals towards a future where BIPV is not merely an option but also an integral element of architectural vocabulary.

May this book inspire, inform and stimulate a new wave of BIPV implementations across the global architectural landscape.

Francesco Frontini and Helen Rose Wilson
IEA PVPS Task 15 co-managers

Acknowledgments

This book has been developed thanks to the collaboration and contributions of many experts linked to IEA PVPS Task 15. We extend our deepest gratitude to all of them.

Remarkable collaboration has been throughout the development of Chapter 5; the assembly of technical information for the BIPV case studies was an immense task that could not have been accomplished without the collective efforts of our esteemed contributors. We thank Anna Svensson (Soltech Energy, Sweden), Isa Zanetti (SUPSI, Switzerland), Angel Gallego, Elena Rico and Diego Cuevas (Onyx Solar, Spain), Jennifer Adami (EURAC, Italy), Enza Tersigni and Carolina Girardi (UNINA, Italy), Sydney Browne (Diamond Schmitt Architects, Canada), Veronique Delisle (CanmetENERGY, NRCAN), Karin Kappel (Solar City, Denmark), Hisashi Ishi (Lixil, Japan), Veronika Shabunko (SERIS, Singapore), Rickard Nygren (White Arkitekter, Sweden), Jose Maria Vega de Seoane (Tecnalia, Spain), Maximilian Lugmair, David Rinnerthaler and Michael Grobbauer (FH Salzburg, Austria) and Dieter Moor (Ertex Solar, Austria), for their support in gathering the needed documental and graphic material and for their invaluable insights and dedication.

We also extend our sincere thanks to the companies and research institutions that generously shared the technical information of their BIPV projects. Their openness and collaboration have been fundamental to this comprehensive publication.

Developing the technical BIPV drawings in Chapter 5 required a harmonious blend of technical rigour and creativity. Our heartfelt appreciation goes to Marley Dowling, Sara Eskandar, Amy Markwart and Patrick Angkiriwang (University of Waterloo), Antonio Rios Calderon and Lorenzo Olivieri (Universidad Politécnica de Madrid), and Emily Zhu and Veronique Delisle (CanmetENERGY, NRCAN, Canada) for bringing clarity and detail to our illustrations.

We also thank for their insights and discussions Alessandra Scognamiglio (ENEA, Italy), Francesca Tilli (GSE, Italy), Karl Viriden (Viridén+Partner, Switzerland) and Hua Ge (Concordia University, Canada).

Special acknowledgement is given to Helen Rose Wilson for reviewing this book and her exemplary leadership and vision as co-task manager of IEA PVPS Task 15, continuing the former IEA PVPS Task 15 leaders Johannes Eisenlohr and Peter Illich's support of the guidebook development. Their guidance has been instrumental in bringing this project to fruition.

Our sincere thanks to the IEA PVPS Executive Committee members Yves Poissant (CanmetENERGY, NRCAN), Hubert Fechner (Fachhochschule Technikum Wien, Austria), Stefan Oberholzer (Swiss Federal Office of Energy, Switzerland) and the IEA PVPS Chair Daniel Mugnier (Planair, France) for the comprehensive review of this publication.

The editors

Contributors

Martín Chivelet, Nuria
Department of Energy
CIEMAT Research Centre
Madrid, Spain

Kapsis, Costa (Konstantinos)
Department of Civil and Environmental Engineering
Faculty of Engineering
University of Waterloo
Waterloo, Ontario, Canada

Frontini, Francesco
Department for Environment Construction and Design (DACD)
University of Applied Sciences and Arts of Southern Switzerland (SUPSI)
Mendrisio, Switzerland

Reijenga, Tjerk
BEAR-iD Architects and Urban Planners
Gouda, The Netherlands

Bonomo, Pierluigi
Department for Environment Construction and Design (DACD)
University of Applied Sciences and Arts of Southern Switzerland (SUPSI)
Mendrisio, Switzerland

Schneider, Astrid
Institute of Architecture and Design
Faculty of Architecture and Planning
Technische Universität Wien
Vienna, Austria

Marcos Castro, Ana
Department of Energy
CIEMAT Research Centre
Madrid, Spain

Stølen, Reidar
RISE Fire Research
Trondheim, Norway

Fjellgaard Mikalsen, Ragni
RISE Fire Research AS
Trondheim, Norway

Wilson, Helen Rose
Fraunhofer Institute for Solar Energy Systems ISE
Freiburg i. Br., Germany

Eder, Gabriele
OFI Österreichisches Forschungsinstitut für Chemie und Technik
Vienna, Austria

Astigarraga, Alexander
Institute for Renewable Energy
Eurac Research
Bolzano, Italy

Ishii, Hisashi
LIXIL Corporation Technology Research Institute
Shinagawa district (ward), Tokyo, Japan

Boddaert, Simon
CSTB Scientific and Technical Centre for Building
Sophia Antipolis, France

Parolini, Fabio
Department for Environment Construction and Design (DACD)
University of Applied Sciences and Arts of Southern Switzerland (SUPSI)
Mendrisio, Switzerland

Rebecca Yang
Solar Energy Application Lab
School of Property, Construction and Project Management
RMIT University
Melbourne, Australia

Lugmair, Maximilian
Salzburg University of Applied Sciences
Salzburg, Austria

Wijeratne, Pabasara
Solar Energy Application Lab
School of Property, Construction and Project Management
RMIT University
Melbourne, Australia

Weerasinghe, Nilmini
Solar Energy Application Lab
School of Property, Construction and Project Management
RMIT University
Melbourne, Australia

Kovacs, Peter
RISE Research Institutes of Sweden
Borås, Sweden

CHAPTER 1

Introduction

Nuria Martín Chivelet

Building-Integrated Photovoltaics (BIPV) is an innovative technology resulting from developing photovoltaic (PV) modules with building envelope construction functions. This means that BIPV modules meet both the electrical and construction requirements according to the role they are to develop as elements of roofing, façades, or external structures. New photovoltaic technologies, materials, and technical progress in the construction industry have permitted the development of BIPV innovative solutions for products and systems adapted to new construction and architectural requirements.

Replacing traditional building products with their BIPV equivalents shows an outstanding potential to contribute to the current and future cities' decarbonisation. However, the use of BIPV in new buildings and renovations is still underrated, partly because there is a lack of technical information addressed to the stakeholders, who need knowledge and guidance to design successful BIPV systems. This guidebook intends to contribute to filling this gap.

The guidebook reviews the main characteristics of BIPV products and systems, as well as their technical and aesthetical requirements and possibilities. Moreover, it aims to help in the decision-making process to achieve efficient and resilient BIPV projects. An attractive highlight is a collection of 24 outstanding international BIPV case studies covering a variety of products, architectural applications, and climates to illustrate the challenges and technical solutions of the BIPV technology; technical drawings and pictures are part of the information tools this guide offers to the reader. The book is structured into six chapters, and the main content is summarised as follows.

Chapter 2 reviews the main energy characteristics and requirements for BIPV modules as electrical and construction products. BIPV module efficiency is a key electrical property, which is PV-technology dependent but also a function of BIPV design features, such as colour or transparency characteristics. At the system level, suitable designs can enhance BIPV systems' electrical performance and optimise the direct self-consumption of locally generated PV electricity. Thermal performance, solar heat gain, daylighting water control, and acoustic performance are some construction requirements analysed before addressing durability, reliability, and safety aspects, which include electrical, mechanical, and fire safety requirements. Finally, aesthetical requirements and recommendations are included to help achieve visually pleasant BIPV projects.

Chapter 3 describes the BIPV products from their components to their final layout as part of the building envelope system. The different BIPV module technologies can vary based on the photovoltaic cell materials and other module components, such as the front and back covers or the encapsulant, which can modify the final layout and characteristics of the product. The manufacturing techniques also allow BIPV modules to fulfill different construction and architectural requirements. The chapter shows the BIPV product design possibilities, challenges, and development trends for its integration into roofs, façades, and shading devices.

Chapter 4 aims to help in the decision-making process of BIPV systems design. Climate, location, orientation, and shading define the base to determine the PV generation potential of the building, with the help of available simulation tools. Then, BIPV solutions that the building offers should combine with energy conservation and energy efficiency measures and optimise the direct self-consumption when applicable. Local regulations and related business models can be decisive in the final electrical and constructive design, also conditioned by the context.

Chapter 5 addresses the practical technical details for developing constructive BIPV solutions. Technical notes and drawings for different types of roofings, façades, and external devices are

DOI: 10.1201/9781003432241-1

meant to serve as a guide for architects and engineers in the design process of BIPV systems. A selection of 24 outstanding BIPV case studies, with detailed technical descriptions, pictures, and technical drawings, is intended to be a motivation source for developing successful BIPV projects.

Chapter 6 collects the most important aspects to consider in an efficient operation and maintenance (O&M) plan for BIPV systems, which aims at improving the overall quality management of the BIPV system's life cycle, mitigating potential risks, and reducing maintenance costs. The main components of an O&M plan include safety aspects, data monitoring, and performance indicators. Maintenance key aspects include the needs of the different BIPV system components, and some additional considerations refer to energy storage systems.

Annex I contains a list of the main symbols and acronyms used in the book. Annex II includes the references of all the standards that are named in the book, alphabetically ordered, since they are not included in the reference section under each chapter.

CHAPTER 2

BIPV performance requirements

Nuria Martín Chivelet, Francesco Frontini, Tjerk Reijenga, Reidar Stølen, Helen Rose Wilson, Gabriele Eder, Ragni Fjellgaard Mikalsen, Fabio Parolini, Hisashi Ishii, Simon Boddaert, Rebecca Yang, Ana Marcos Castro and Peter Kovacs.

2.1 Introduction

Over the past three decades, the International Electrotechnical Commission (IEC) and other organisations have established standards for photovoltaic (PV) modules and systems addressing their electrical performance, stability, and safety. Similarly, the International Organisation for Standardisation (ISO) has established design and manufacturing specifications and processes covering every part of a construction project. Those standards apply to BIPV modules and systems from the electrical and construction perspectives. Combining both aspects, IEC 63092 (parts 1 and 2) is the first international standard for BIPV products and systems. Under IEA PVPS Task 15, several publications have addressed some different aspects related to performance and safety needed for the development of BIPV standards, e.g., [1,2], and to design possibilities and energy performance modelling of BIPV, e.g., [3,4]. This chapter addresses the main performance requirements for BIPV products and systems, considering their roles as energy generators and construction materials in buildings.

The performance requirements of BIPV modules and systems considered in this chapter refer to the following main properties: electricity generation, thermal, solar, and optical performance, water control and weatherproofing, acoustic insulation, durability, reliability, aesthetics, and mechanical, electrical, and fire safety.

2.2 Electricity generation

From an electrical standpoint, a BIPV system is very similar to a "standard" PV system. However, the electrical performance of a BIPV system can significantly deviate from that of a standard PV due to the unique constraints imposed by the built environment. Key differences include susceptibility to partial shading, irregular irradiance patterns due to nearby structures or reflections, and reduced back ventilation leading to higher module temperatures. Extreme tilt angles (high or low) may also affect angle-dependent and/or soiling losses. BIPV designers should balance architectural constraints and electrical efficiency, aiming to minimise power losses associated with the specific conditions encountered in BIPV installations.

In the next paragraphs, the electrical requirements will be presented and discussed, starting with an examination of BIPV modules and progressing towards their integration into the overall building energy system.

2.2.1 BIPV module efficiency

The efficiency of a PV module (η) is defined as the ratio of the output power per unit of area to the incident irradiance. As the energy yield is highly dependent on the exposed conditions making it difficult to define the efficiency of a module, the international standards community and industry have agreed to evaluate the nominal efficiency of PV modules under standard test conditions (STC), which are irradiance 1 kW/m², cell temperature of 25 °C, and AM1.5 global solar spectrum as stated in the standard IEC 60904-1 (Equation 2.1):

$$\eta_{STC} = \left(P_{max} / A_m\right) / G_{STC} \qquad (2.1)$$

where P_{max} is the module's maximum power at STC, A_m is its total area (defined by its outer edges), and G_{STC} is 1 kW/m². Efficiency is commonly expressed as a percentage; for example, a nominal PV module efficiency of 18% implies that, under STC,

DOI: 10.1201/9781003432241-2

Table 2.1 Typical efficiencies at STC of most common PV module technologies.

Photovoltaic technology	mono-crystalline silicon	multi-crystalline silicon	silicon hetero-junction	amorphous silicon	cadmium telluride	copper Indium (gallium) selenide
Acronym	m-Si	mc-Si	SHJ	a-Si	CdTe	CI(G)S
Efficiency	19%–22%	18%–20%	18%–22%	6%–8%	17%–19%	16%–18%

its maximum output power per square metre is: 18% × 1000 W/m² = 180 W/m². Similarly, a PV module with a surface area of 3 m² and efficiency of 18% will yield maximum output power under STC equal to: 18% × 3 m² × 1000 W/m² = 540 W per PV module.

When calculating the BIPV module efficiency, the total surface area of the module should be taken into account (i.e., covered and not covered with solar cells), as defined by its outer edges, including the frame, if any. When compared to regular PV, some BIPV designs reduce the number of cells that cover the surface area of the module to increase its visible transmittance (see Section 2.3) or to create a particular pattern. Another common difference between BIPV and conventional PV concerns the module frame, which may either differ from standard products or be omitted completely.

Special designs to fulfil the thermal, solar, and optical requirements for building applications can lead to BIPV modules with lower electrical efficiencies than standard PV modules. Table 2.1 shows typical ranges of electrical efficiencies at STC for most common opaque modules; as PV cell and module technologies continue to develop, efficiencies will continue to increase, exceeding the values provided in Table 2.1.

2.2.2 Colour impact on energy performance

BIPV modules can show different colours and textures to enhance the aesthetics of the building. Literature referring to innovations and the market for coloured BIPV [3] has shown the technological progress of BIPV technologies to respond to architectural demands. Altering the colour of a standard module may lead to a reduction in its electrical efficiency. This occurs if the transmission of solar radiation in the spectral range absorbed by the PV cell is significantly reduced by the colouring medium as it is the case with coloured pigments (absorptive colours). However, when structural colours are used, the performance loss is lower. In this case, the coloured appearance is achieved by spectrally selective reflectance in only a narrow spectral range by applying interference coatings onto suitably treated surfaces; it is possible to combine vivid colours that remain stable over a wide range of viewing angles with low power losses of less than 10% in relative terms [5,6].

The colouring can be achieved by the PV cell itself, the encapsulant, interlayers, or the front or back covers. The changes affecting the materials located behind the cell involve small or insignificant efficiency losses. By contrast, changes in the materials in front of the cell may cause significant losses. In these cases, the losses are very colour-dependent (Figure 2.1).

Some investigations on coloured cells with dedicated coatings are ongoing. Based on the principle of anti-reflection coatings (AR coatings), the colour change is mainly due to the variation of coating thickness and typically lies in the range of 8–20%. Silicon nitride or titanium oxide are commonly used to change the coloured appearance of cells.

Figure 2.1 Coloured demonstration façade at CIEMAT, Spain. In this example, green and terracotta BIPV modules reduce the efficiency by 4% and 15%, respectively, when compared to the anthracite colour. Ref: RINGS-BIPV project, 2023. Source: N. Martín-Chivelet, CIEMAT.

2.2.3 BIPV system's efficiency and performance ratio

Solar irradiation is the dominant parameter affecting PV electricity generation. It can be assumed that PV power output is primarily proportional to incident in-plane irradiation. However, secondary factors such as module cell temperature also impact BIPV system performance. The challenge with BIPV, especially in vertical façade applications, lies in the complex nature of irradiation calculations. BIPV façade systems often experience uneven irradiance due to varying module positions and shading, even within the same façade plane. This makes a single measurement or calculation for the entire system insufficient. Instead of relying on the simple view factor methods used in conventional PV system calculations, it is advisable to conduct a detailed, time-series shading analysis unless partial shading is absent. Some available software tools can assist in this task. A comprehensive review is available in [4] and [7]. For more accurate estimations and measurements of actual irradiance, analysing the system in segments is recommended.

The increase of module cell temperature negatively affects the output power of a PV module. As a first-order approximation, temperature losses can be written as:

$$L_{temp} = -\gamma \cdot \left(T_c - T_{c,ref}\right) \tag{2.2}$$

where T_c denotes cell temperature (the subscript *ref* stands for reference conditions, e.g., 25°C under STC), and γ is the power temperature coefficient, representing the change of the maximum power of a PV device per unit change of temperature; γ is technology-dependent, with crystalline silicon being the most sensitive, displaying an efficiency decrease of around 0.45% for every degree increase of temperature. Typical values for each technology range from 0.25–0.28% for amorphous silicon and cadmium telluride to 0.36–0.46% for CIGS and crystalline silicon. More details can be found in [8].

In order to assess the performance of different BIPV systems, the Performance Ratio (*PR*) can be used. The *PR* is the ratio of the actual over the theoretically possible energy yield of an ideal installation with neither environmental nor system-based losses, under the same irradiation conditions, as stated in IEC 61724. Consistently, it is the result of multiplying all the efficiency-reducing factors corresponding to each loss coefficient L_i:

$$PR = \left(1 - L_{temp}\right) \cdot \left(1 - L_{soiling}\right) \cdot \left(1 - L_{wiring}\right) \cdot \left(1 - L_{inverter}\right) \cdot \ldots \cdot \left(1 - L_i\right) \tag{2.3}$$

As in other PV systems, there may be losses related to temperature increase, solar spectral changes, optical reflections, surface soiling, wiring, losses in power electronics (e.g., maximum power point tracking, DC-to-AC conversion), and other PV system component inefficiencies. Some losses are site-dependent, because of the meteorological conditions, but others are design-related and should be reduced as much as possible through appropriate design, mounting, and maintenance measures. It should be noted that the losses due to optical reflections are often higher for BIPV than for ground-mounted PV, as the tilt angle and orientation are determined primarily by the building envelope, such that the solar radiation is often incident at more oblique angles [9]. The *PR* can be used to compare BIPV installations at different locations around the globe.

The performance ratio can be calculated over a given reporting period. Although daily values can give more detailed information about possible issues, monthly and annual values are usually used to assess the performance of a BIPV system. Values above 74% are achievable with good system designs [10]. However, it is difficult to establish a typical BIPV *PR* range since it is strongly dependent on the climate and the tilt angle. Moreover, PV technology affects the *PR* because of the different temperature sensitivity of the materials [11].

2.2.4 Electrical design of a BIPV system

One of the main constraints when designing a BIPV system is the non-homogeneity of solar irradiance over the entirety of the modules that make up the system, and the more frequent partial shading than in conventional ground-mounted or rooftop PV systems. Power loss from partial shading is often assumed to be proportional to the shaded area of the PV generator over time [12], although shading losses are usually higher and dependent on the electrical design of the PV modules and PV system. Some authors assume that shading affects only the direct and circumsolar components of the solar irradiance when calculating the hourly shading effect over one year and obtaining the annual shading factor F_s, which is useful to estimate the effective solar irradiation that reaches the PV generator.

Apart from power losses, partial shading under high irradiance may also cause damage in PV modules due to localised overheating (generating hotspots) if the system is not designed appropriately. BIPV module and system design strategies have been developed to avoid or reduce the partial-shading effect, such as including dummy cells where the shading is expected or increasing the perimeter margin around cells in the modules to prevent shading from the frame or the mounting structure. If partial shading of active parts of the PV modules cannot be avoided, which is the case for many BIPV systems, the electrical design of the system has to be adapted to the specific situation.

Under specific undesirable conditions, shaded cells can be reverse-biased, dissipating the power generated by the other cells of the string. To avoid such hotspots, all modules feature bypass diodes that can short-circuit complete modules or sub-strings of the module. For crystalline silicon cells, for example, typically sub-strings of 20–24 cells are protected by one bypass diode. As partial shading occurs more frequently in many BIPV systems than in ground-mounted systems, these bypass diodes have to be chosen carefully to ensure a long lifetime also under frequent partial shading conditions.

In general, the output power of thin-film PV modules is less sensitive to partial shading than are crystalline silicon ones, due to the PV cell geometry and configuration within the module [13]. In addition, the currents are usually lower in thin-film modules, and this makes reverse-biased thermal effects less harmful. However, under high irradiance conditions and extreme partial shading, cell defects could occur in any PV cell technology. PV module designs with divided crystalline PV cells can help reduce these effects.

The presence of partial shading or, more generally, non-uniform solar irradiance across modules, makes the electrical design a key aspect to ensure the optimal performance of a BIPV system. The main reason is that the series connection of PV modules exposed to different irradiance levels can reduce the string current to the lowest module current, which corresponds to the lowest irradiance. Equivalently, parallel connections of modules or strings with different individual voltages need to be handled with care. In general, all strings connected in parallel should have the same voltage level under all occurring operating conditions and, in some situations, fuses or string diodes need to be used to avoid reverse bias of parallel strings.

Inverters are the electronic devices used in PV systems to convert direct current (DC) from the PV modules to alternate current (AC), to use it directly at the building location or feed it into the grid. Selecting a central inverter configuration, which means connecting the whole PV generator (all the modules) to one inverter, is not always advisable for BIPV, even with appropriate bypass diodes. The reason is the frequent non-homogeneity of solar irradiance over the totality of modules described previously. Module-level inverters (microinverters), which are small inverters located behind the modules to connect them individually or in small groups directly to the AC grid, can be a good option for systems with strongly varying operating conditions. An intermediate solution between central and module-level converters is using string or multi-string inverters, which are designed to be connected to strings of modules with the same operating conditions.

As a general rule, if no partial shading is foreseen and all of the modules are in the same plane, the central inverter could be the best option (the most cost-effective solution with high efficiency and durability). Otherwise, the PV system should be electrically divided into segments corresponding to different irradiance conditions, and turned into multi-string or string inverters, or even microinverters. Another option is to use DC-DC optimisers to regulate the output, as will be explained in Chapter 4.

2.2.5 Photovoltaic self-consumption

To enhance the profitability and sustainability of BIPV or BAPV (building-attached photovoltaic) projects, it is essential to maximise the local utilisation of the electricity they generate. This concept is commonly referred to as 'self-consumption,' wherein individuals or businesses harness the energy generated by their own photovoltaic systems that are strategically positioned close to the point of energy consumption. Self-consumption of photovoltaic energy, especially from residential and commercial PV systems, stands as a pivotal catalyst for the integration of PV technology into building infrastructure. Various models, including individual, collective, and energy community models, have been adopted in numerous countries, rendering self-consumption financially attractive and reducing the strain on public electricity distribution grids. This approach facilitates the immediate consumption of (BI)PV-generated electricity, resulting in cost savings on electricity bills and achieving a more harmonious balance between supply and demand, thereby benefiting grid operators. Two indices are commonly used to describe the use of the self-produced electricity: the self-consumption index (*SCI*) is the share of the total PV production directly consumed by the system owner, while the self-sufficiency index (*SSI*) indicates the percentage of the energy consumed that comes from the PV energy locally generated:

$$SCI\,(\%) = 100 \times \text{PV electricity consumed/PV electricity generated} \tag{2.4a}$$

$$SSI\,(\%) = 100 \times \text{PV electricity consumed/electricity consumed} \tag{2.4b}$$

SSI, which becomes especially important for fluctuating high prices of electricity, typically oscillates between 20% and 25% in residential or commercial buildings. Exceeding 40% is difficult without the support of energy storage [14]. Another way of increasing self-sufficiency is demand-side management (DSM), a strategy to help consumers to optimise their energy usage by adjusting the amount and timing of electricity use [15]. DSM can also be combined with electricity storage if necessary [16].

It is important to note that the optimised BIPV system design can significantly improve the SCI and the SSI in commercial and office buildings. The system should ensure that the BIPV modules occupy those parts of the building envelope that lead to the best fit between the PV generation and the building load profiles [17]. A good match between them can easily lead to 100% direct self-consumption indexes and increase self-sufficiency.

Figure 2.2 shows an example of electricity consumption and generation in a multi-family house in Chiasso (in southern Switzerland) during typical summer and winter days. The lighter grey area in both graphs represents the electricity generated by the PV systems not consumed by the building and, therefore, fed into the grid. Appropriate DSM or electrical storage can increase the *SCI* of the system. More information can be found in [18].

With DSM techniques and, when necessary, energy storage, self-consumption can be significantly enhanced by aligning energy demand more effectively with PV electricity generation. DSM is designed to establish an equilibrium between electricity demand and generation over specific time intervals, be it by hour, day, week, or year.

However, challenges do arise, particularly when dealing with less-than-ideal electricity generation profiles stemming from fluctuating sunlight conditions and unpredictable energy consumption patterns. In response to these challenges, contemporary PV inverters, tailored for self-consumption, often incorporate advanced smart energy management systems. These systems aim to optimise the utilisation of PV generation, align it with consumption patterns, and, in some cases, incorporate battery storage and electric vehicle (EV) charging solutions, ultimately promoting greater self-sufficiency and efficiency.

2.3 Thermal performance and solar gain

As part of the building envelope, a BIPV system must protect the building from outdoor conditions such as rain, temperature fluctuations, wind, and other natural phenomena. In this chapter we will discuss the temperature behaviour and thermal performance of BIPV systems.

Thermal transmittance

In applications such as BIPV ventilated façades or discontinuous roofing, the thermal insulation of the building is mostly determined by the properties of the insulation material directly attached to the building. By contrast, in glazing applications, such as curtain walls, windows, atriums, and skylights, the BIPV module design plays a major role in the thermal transmittance (characterised by the U value) of the building envelope.

Since the U value of a building component is dependent on the boundary conditions, the international standards for glass in buildings have established a set of test conditions for outdoor and indoor temperatures, outdoor and indoor convective heat

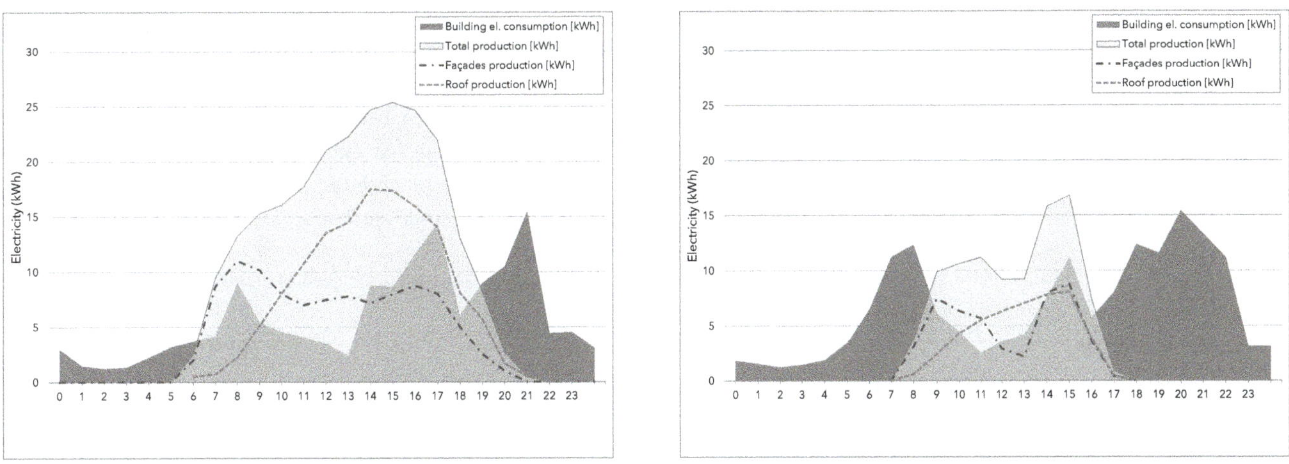

Figure 2.2 Example of electricity consumption and generation daily profiles for an apartment block in Chiasso (Switzerland) during summer (left) and winter (right) [18]. The multi-storey building was renovated in 2012; all four façades integrate BIPV modules with a total installed power of 52 kW.

transfer coefficients, and solar irradiance (G = 0). Similar to other construction products, BIPV modules have been tested to determine their U values, generally with guarded hot-plate equipment, by calculation from dimensions and material properties, or by heat flow metre methods. IEC 63092 refers to three glass-in-building procedures described in ISO 10291, ISO 10292, and ISO 10293, respectively, to determine the U value of BIPV modules. According to IEC 63092, the aforementioned standards apply to any BIPV module having at least one glass pane. This is the case for all BIPV modules with a front cover made of glass, representing the vast majority of BIPV modules.

According to [19], the U values of a BIPV module and the equivalent non-PV glazing are similar under controlled boundary conditions. Simulations reported in [20] indicate that the change in U value for vertically installed insulating glazing is small (< 0.05 W/(m^2 K)). In general, under operating conditions, the U value of a BIPV module may slightly change as a function of the solar irradiance, mainly due to the irradiance effect on the energy conversion of the photovoltaic device and its temperature [21].

2.3.1 BIPV module temperature

The temperature of BIPV modules affects their thermal characteristics as construction elements and their electrical performance, and can also impact their durability (e.g., longevity and degradation of certain polymer components). Therefore, modelling BIPV module temperature is essential not only for accurately predicting thermal, solar, and electrical properties but also for assessing the material's overall performance and durability. Steady-state models are often used for stable conditions due to their simplicity, assuming an immediate response of the module temperature to changes in irradiance and wind speed [22]. One example is King's or the Sandia model [23]:

$$T_{\text{back}} = Ge^{(a+b.w_s)} + T_a \qquad (2.5)$$

where T_{back} is the PV module back-surface temperature, G is the in-plane solar irradiance (W/m^2), T_a is the ambient air temperature, and w_s is the wind speed (m/s). Parameters a and b depend on the module construction and materials as well as on the mounting configuration of the module [24] (see Table 2.2). The difference between module cell temperature under equation (2.2) and module back-surface temperature in equation (2.5) is a function of the solar irradiance, and it is approximated as follows:

$$T_c - T_{\text{back}} = \frac{G}{G_0} \cdot \Delta T \qquad (2.6)$$

where G_0 is the reference solar irradiance (i.e.,1000 W/m^2) and ΔT is the temperature difference between the cell and the module back-surface at that irradiance. Table 2.2 provides representative values for the coefficients a, b, and ΔT for various BIPV configurations.

Other commonly used steady-state models are those included in the IEC PV standards. One of them is based on the simple former Ross model [25] and defines the nominal operating cell temperature (NOCT) as the characteristic parameter to fit in each case. The other is based on the Faiman model [24] to determine the nominal module operating temperature (NMOT). The parameters of each model are fitted by using experimental data obtained with a set of standard test boundary conditions. The effect of back ventilation on the PV module temperature has been reported in the literature. Moreover, some authors have fitted the NOCT parameters for different BIPV modules and system configurations [26–28].

The simplified analytical models described might give acceptable results for a rough forecast of PV energy over long periods (e.g., one year). The recommendation is to fit the corresponding parameters by linear regression of measured temperature data to the specific model or, at least, to adopt the parameters recommended by the model developers. If higher accuracy is needed, considering the energy balance equations for each module layer could be the solution [21].

Based on a year-long monitoring test conducted across seven countries within the framework of an IEA Task 15 round-robin initiative [29], it was observed that the module temperatures in BIPV ventilated façade mock-ups typically fluctuate between the

Table 2.2 Typical values recommended by [8] for coefficients a and b to estimate PV module back-surface temperature (Equation 2.5), and approximate temperature difference ΔT between that back-surface and PV cell.

Type of PV application	a	b [(m/s)$^{-1}$]	ΔT (K)
Open rack PV, BAPV, or BIPV with good rear ventilation	−3.47	−0.0594	3
BAPV or BIPV with medium rear ventilation	−2.98	−0.0471	1
BAPV or BIPV with poor rear ventilation	−2.81	−0.0455	0
Double-glazed BIPV window with one low-emissivity coating (surface 2 or 3)	−2.85	−0.0351	9
Triple-glazed BIPV window with two low-emissivity coatings (surfaces 2 and 4, or 3 and 5)	−2.88	−0.0319	11

ambient temperature and approximately 60°C. Poor ventilation and higher irradiance at other possible tilt angles might increase the maximum value by some degrees. Equations (2.5), (2.6), and Table 2.2 help to estimate the module cell temperatures at any specified boundary conditions. As an example, Gok et al reported 15 °C difference between ventilated and insulated modules [60].

2.3.2 Solar heat gain coefficient

The solar heat gain coefficient (*SHGC*), also known as g value, solar factor, or total solar energy transmittance, characterises the solar-control ability of architectural glazing, including semi-transparent BIPV modules [30]. It is defined as the fraction of incident solar radiation that passes through the module to the inside of a building, both directly by transmission (characterised by the solar transmittance, T_{sol}) and indirectly (characterised by the secondary internal heat transfer factor, q_i) after absorption and inward re-emission of heat. Glazing products with low g values reduce the cooling load in summer, while large g values lead to a reduced heating load in winter.

$$SHGC = T_{sol} + q_i \tag{2.7}$$

In glazing units consisting of one or more flat, parallel layers, the *SHGC* for each optically different area can be determined separately by calculations based on spectral measurements, as described in two ISO standards (ISO 9050 and ISO 15099). As noted in the European standard EN 410, the *SHGC* can then be area-weighted to obtain the *SHGC* for the complete unit. This calculation-based approach is valid for all semi-transparent BIPV modules, whereby the PV cells, the transparent areas between them, and the cell connectors are the "optically different areas."

The *SHGC* can be determined experimentally by the calorimetric approaches described in ISO 19467, considering that the evaluated area of the BIPV module should be optically representative of the total area of the module [30] (Figure 2.3). Any *SHGC* calculation or measurement should consider the electrical state of the BIPV module, from open-circuit (no absorbed energy is converted into electricity) to the maximum power point (maximum electricity extraction), which affects the *SHGC*, as explained in [31, 32]. The *SHGC* value reduces when electricity is extracted from the module.

Figure 2.3 Indoor measurement of the *SHGC* of a BIPV insulated glazing unit operating under maximum power point, at the Centre for Zero Energy Building Studies. Photo credits: Costa Kapsis, University of Waterloo.

2.4 Daylighting and visual comfort

BIPV modules can be semi-transparent and let daylight enter the building, contributing to the energy savings of the building and enhancing the visual integration in the envelope. In principle, all PV technologies can be designed with a certain degree of transparency. In crystalline silicon BIPV modules, light transmission can occur through the gaps between cells. As a result, light through them creates a non-uniform luminance pattern due to the combination of opaque PV cells and transparent gaps (Figure 2.4). By contrast, semi-transparent thin-film BIPV modules provide uniform luminance distributions indoors and low obstruction of the outdoor view (Figure 2.5).

The colour rendering index (*CRI*) of the transmitted light through glazing units, including semi-transparent BIPV modules, characterises the ability of the light to reproduce colours faithfully when compared to standard natural light. The CIE method [33] for calculating the colour rendering is recommended for use in semi-transparent BIPV façades and shading devices. The international standard ISO 8995-1 specifies the lighting requirements for indoor workplaces and for people to perform visual tasks efficiently, comfortably, and safely. It establishes specific ranges of values for the *CRI*: ideally, it should be above 90, although values above 80 could be acceptable depending on the use. Poorer colour rendering may cause discomfort.

Figure 2.4 Semi-transparent BIPV skylight using c-Si solar cells at a school, in Courbevoie, France. Photo credits: MTECH & CSTB.

In general, semi-transparent PV modules have high colour rendering indexes. However, coloured BIPV designs achieved by modifying the encapsulant or the front or backsheets reduce the colour rendering quality, although colours may benefit aesthetics.

2.4.1 Water control and weatherproofing

Integrating PV modules in traditional roofs is a widespread practice in renovation and energy-efficiency improvement of buildings. Different research projects, publications, and real projects have demonstrated that several technologies available in the market offer a variety of dimensions, shapes, and colours, offering aesthetically attractive solutions (e.g., [34]). Together with the performance aspects presented earlier, the manufacturers and installers must consider the thermo-hygrometric properties and the reduction of water leakage or condensation in the area around and underneath the modules. These aspects are particularly relevant in discontinuous roofs but also apply to façade elements when water tightness is required. Rain penetration into the building envelope can create problems affecting the durability of building materials, such as material degradation, mould growth, and wood decay. Rainwater can reach inner roof structures through the areas where the roofing underlay is fastened by nails and staples [35].

Some methodologies for the evaluation of discontinuous roofing water resistance have been tested. These experiments are based on simulating combined forces and elements such as wind, rain, and flowing water. These forces are applied on a roof specimen, e.g., the French Scientific and Technical Building Centre, CSTB, uses a large wind tunnel to assess BIPV roof mock-ups according to different exposure situations, varying wind speed, and rainfall rate [36], with different roof pitches and roof orientation. Some investigations are currently in progress to merge CSTB work with the testing facility proposed by Norwegian University of Science and Technology NTNU, based on the use of a rain and wind box to assess the rate of water penetration under several different weather conditions [35]. In both investigations, CEN/TR 15601 concerns the method to test the resistance to wind-

Figure 2.5 Retrofit of a skylight made of semi-transparent a-Si BIPV modules, in the historic Béjar Market in Salamanca, Spain. Photo credits: Onyx Solar.

driven rain of roof coverings with discontinuously laid small elements.

The principle of the roof waterproofing test is based on spraying a certain amount of water at different air pressure differences at different angles and defined conditions onto the outer surface of a roof sample to observe whether water leakage occurs.

The limit on the amount of water allowed to penetrate the roof system is seldom adequately specified for building envelope systems. For BIPV systems intended for roof integration, this aspect is addressed in the European standard EN 50583-2 "Photovoltaics in buildings. Part 2: BIPV systems," in annex A "Resistance to wind-driven rain of BIPV roof coverings with discontinuously laid elements – test method" [37].

A water collector shall be provided, capable of recording the amount of leakage water during any pressure step in the test. The reference leakage rate is (10 g/m2)/5 min, 5 min being the duration of a single test step in the sub-test. The cases where leakage occurs via fine spray and underside wetting are considered too severe for the application.

2.5 Acoustic insulation

The acoustic performance of a BIPV glazing façade is related to the mass and composition of the glasses, as well as the air permeability of the joints in the enclosure. Insulation to external noise can be improved by installing fillings of some noise-absorbent material and ensuring maximum tightness to air entrance (e.g., see Figure 3.17 in [38]). The thickness of the glass, the infiltration of air through the carpentry, and the glazing type are the main parameters that control the acoustic performance of a BIPV glazed façade. Minor factors are the type of warm edge in multi-pane insulating glass units (IGUs), the thickness and type of gas in the air chamber, and glasses' size.

In general, BIPV modules increase the acoustic insulation performance, compared to single glazing, because they are based on laminated glass. Moreover, the resonance effect can be reduced if the two panes of glass of the BIPV module are chosen to have different thicknesses. Also, overhangs and horizontal louvre-shaped modules for solar shading may improve acoustic insulation performance to some extent.

2.6 Durability, reliability, and safety

2.6.1 Thermal stress resistance

Many different materials form the BIPV module and its mounting structure, such as aluminium alloys and steel for frame and support structure, semiconductors and metallic contacts and wires for PV cells, polymers for encapsulant, and other materials for junction box, sealing, and gaskets to ensure water tightness and air tightness.

Widely different linear expansion coefficients of the materials and diverse manufacturing processes may lead to significant variations in expansion and contraction behaviour when substantial temperature fluctuations occur. This phenomenon necessitates careful consideration in the design of new modules and material selection. For this purpose, IEC 61215-2 includes two cycle tests:

1. Temperature cycle test to determine the ability of the module to withstand thermal mismatch, fatigue, and other stresses caused by repeated temperature changes. In the temperature cycling test (IEC 61215-2 MQT 11), the PV module is exposed 200 times to temperature cycles between -40°C and +85°C.
2. Humidity freeze test (IEC 61215-2 MQT 12) to determine the ability of the module to withstand combined high temperature and humidity followed by sub-zero temperatures.

After both tests, a wet-leakage test (IEC 61730-2 MQT17) has to be carried out to ensure the absence of defects and module safety.

2.6.2 Wind load resistance

When designing a BIPV system, it is essential to keep in mind that BIPV replaces traditional building envelope materials. Consequently, ensuring robust mechanical strength is imperative to prevent permanent deformation or damage caused by wind loads on the building. Wind load refers to the forces and pressures the wind exerts on a building, structure, or its components. Extreme weather conditions, such as tornadoes and hurricanes with high-speed winds, can subject BIPV systems to significant stress [39]. Even routine wind gusts resulting from common weather patterns can adversely affect the performance of a BIPV installation. Therefore, designers must prioritise BIPV wind load resistance [40]. Two types of pressure could act on the BIPV systems as a part of the building envelope: (1) positive pressure acting towards the surface and (2) negative pressure (suction) acting away from the surface [41]. Consequently, assessing the wind resistance of PV systems should account for loads in both directions.

For instance, some parts of the building skin may experience negative pressure when air flows along the building's surface layer in a direction opposite to the prevailing wind. This can lead to peeling at the outer corners and upper edges of the building envelope, creating vulnerable areas that require special attention.

Different factors need to be considered when calculating wind pressure, such as building location, height, shape, surrounding topography, terrain category, and shielding multiplier. To meet the requirements of wind load resistance, several calculations must be considered for the BIPV systems, especially when dealing with tall structures or those situated in high-wind-speed regions.

2.6.3 Seismic load

In general, buildings are firmly attached to the ground, so when an earthquake occurs, the response to the vibration waves changes depending on the ground's stiffness and the building's rigidity. The main structure of a building must be sound against the expected vibrations, and the building skins must also be held safely against these vibrations, so that curtain wall frames and glass will not fall off or be damaged. This also applies to BIPV systems.

Curtain wall systems need to be safe against the seismic forces acting from the combined vibration waves (horizontal and vertical directions) when the initial input wave and the eigenvalue match. In addition, they should follow the deformation of the inter-storey (height of a storey) against seismic motion. An earthquake causes a deformation angle between the layers of the main structure, and the curtain wall element must withstand that deformation angle.

Thus, the curtain wall systems can mitigate seismic load if it has a structure that moves when direct seismic loads are applied [38]. Since frames and panel portions behave differently, careful consideration should be given to the wiring and mounting methods when a PV module is designed. This is paramount because wiring may be subjected to displacement due to seismic forces acting on the floor. This has to be taken into account also for BIPV modules installed to cross over the junction between different segments of the curtain wall.

2.6.4 Impact resistance

The impact resistance of PV modules is evaluated through the module breakage test MST 32 included in the standard IEC 61215-2, which aims to minimise the risk of physical injuries if the PV module breaks after installation. However, the standard reports that for building-integrated or overhead applications, additional tests may be required by relevant building codes.

Another impact resistance test included in IEC 61215-2 is the hail impact test MQT 17. In this case, it is also necessary to refer to national and local code requirements in addition to this international standard since the BIPV module's function in the building is relevant. Recent pre-normative activity increases the severity of hail impact tests (larger ice balls and higher velocity) [61].

Therefore, the IEC tests do not sufficiently cover all of the needs for PV in buildings since different safety and impact resistance requirements are needed depending on the architectural function of the building element. As the basic structure of most BIPV modules consists of at least one glass placed in the front position of the module, it is necessary to refer not only to the PV requirements expressed by IEC but also to the standards for glass issued by ISO. ISO 12543 1-6 series applies to laminated glass, defined as an assembly consisting of one sheet of glass with one or more sheets of glass and/or plastic glazing sheet material joined together with one or more interlayers. The standards addressing the impact resistance of laminated safety glass also apply to BIPV laminated glass according to their use in buildings. This means meeting the required performance with respect to the pendulum impact test and corresponding classification. The ISO 29584 standard describes in detail the methods for pendulum impact testing. However, variations may be applied depending on the regulations of each country. In addition, after testing, it should be verified that no visual defects are present, and the electrical performance is unaffected.

The recently reviewed international technical specification ISO/TS 18178 specifies requirements of appearance, durability and safety, test methods, and designation for laminated solar photovoltaic glass for use in buildings. It is applicable to building-integrated photovoltaics (BIPV), although building-attached photovoltaics (BAPV) can also refer to this document.

2.6.5 Mechanical stability and safety

PV modules are mainly made of partially tempered glass or toughened safety glass. Whereas the PV cells of conventional PV modules have been historically laminated between a polymer backsheet and a glass pane, the photovoltaic cells for BIPV modules are more frequently laminated between two glass panes to increase the mechanical resistance and guarantee the required safety. The lamination serves, on the one hand, to embed the solar cells and, on the other hand, to connect the front and back panes of the module. The strength must be equivalent to that of laminated safety glass, so special tests and requirements must be met to ensure adequate mechanical performance. Sufficient strength of PV modules can be proven with project-specific module tests. This is particularly relevant when the modules are installed overhead, as in the case of canopies or skylights, necessitating the use of laminated safety glass and specific tests to assess the residual loads even in case of breakage. IEC 63092-2 specifies that BIPV systems shall withstand the loads, which can be verified either by calculation or by testing.

2.6.6 Thermomechanical degradation

As described in the previous sections, BIPV modules often operate at higher temperatures than PV modules in open rack installations and can also be subjected to static and dynamic mechanical stresses from wind, storm, snow and ice, or hail impact. This combined thermo-mechanical stress on the multi-material composite (PV module) can lead to (1) delamination, (2) glass breakage, or (3) cracks in individual material layers if the connection between the individual layers of the laminate is not strong and/or durable enough. Large differences in the coefficients of thermal expansion in the various spatial directions of the individual layers of the laminate can lead to severe tensions. This effect, intensified by mechanical stress, can lead to delamination and/or material breakage and, consequently, reduced electrical performance and/or safety issues.

The connection between the glass-based PV module and the substructure or the façade system is made either using structural adhesives or, more commonly, using mechanical fixing elements. For safety reasons, all normative requirements that also apply to other construction products (façade elements, roof structures, overhead glazing) must be met. In particular, the standards referring to the test methods for durability of laminated glass and laminated safety glass (ISO 12543-4) and the standard for laminated solar photovoltaic glass for use in buildings (ISO/TS 18178).

2.6.7 Electrical performance degradation

The electrical performance degradation, usually meaning an efficiency reduction, is a critical aspect to consider when evaluating the reliability of BIPV systems. The BIPV exposure to environmental factors, such as solar irradiance, temperature, moisture, mechanical load (wind, snow), soiling, or chemical pollutants can affect their long-term reliability. Additionally, the intrinsic characteristics of the BIPV modules and cells may accelerate that degradation.

In general, a poor design (e.g., in sizing or selection of materials) or bad processing in the manufacturing of PV cells may deteriorate their performance and even cause failures such as hotspots and electric arcs. If the PV cell metallisation or interconnect wiring corrode, the electric resistance of these materials can increase and cause hotspots in the cell. Since the presence of moisture accelerates corrosion, the properties of encapsulant, back and frontsheets, and sealant are critical to protect PV cells from this type of degradation. However, PV materials can also suffer failures and degradation, such as cell cracking, hotspots, light-induced degradation, or potential-induced degradation. These defects and degradation modes are not always easily visible and need detection methods, such as electroluminescence and thermography, as explained in Chapter 6.

Cell cracks can occur during manufacturing, transportation, or installation and can evolve under the operation of the BIPV system, degrading the electrical performance of the modules.

A typical performance degradation occurring in PV modules is light-induced degradation (LID), which arises during the initial stabilisation of a PV module when exposed to light. LID can typically reduce the module power up to 5%, depending on the technology. The stabilised power value is usually given by the BIPV module manufacturer.

Another type of performance degradation is potential-induced degradation (PID). It occurs when the voltage between the PV cells of a module and the ground induces leakage currents that degrade the module's performance over time. The PID effect can lead to significant power losses and is more severe on the negative pole of a string. In the case of BIPV, PID can be exacerbated by factors such as proximity to building materials and potential grounding issues. Mitigating PID in BIPV may involve using PID-resistant materials (e.g., high-volume resistivity encapsulants and modified antireflective coatings), ensuring proper grounding, or adapting the electrical configuration of the BIPV system.

The IEC Technical Specification IEC TS 62804-1 establishes the testing severity representing the minimal stress levels for PID detection in PV or BIPV modules. Some other testing procedures designed to guarantee the long-term reliability and durability of PV modules are included in IEC 61215 and IEC 61730 standards. However, it is recommended that the standards consider further combinations of degradation tests to reproduce the stress of severe climates and circumstances.

2.6.8 Electrical safety

The electrical safety requirements for PV modules in general, and BIPV modules in particular, include avoiding risk to persons due to shock or injury from contact with the electrically live parts of PV modules, as a result of design, construction, or faults caused by the environment or operation. The relevant standard IEC 61730-2 includes the testing procedures to assess and avoid those risks. Examples of electrical shock hazard testing procedures included are the accessibility test, insulation test, or wet leakage current test.

Moreover, the international standard, IEC 60364 for low-voltage electrical installations, applies to BIPV systems. It is an exhaustive document structured into several parts going from the fundamental safety principles, the selection and erection of safe electrical equipment, to safety verification and functional aspects. It also addresses the requirements for special installations or locations, which include solar photovoltaic power supply systems and power supplies for electric vehicles.

Most national codes and regulations implement the safety objectives of IEC 60364 and IEC 61730 on low-voltage electrical installations.

2.6.9 Fire safety

Fire safety is a vital concern in developing solar building envelopes; it is essential to ensure that using PV modules in façades and roofs to replace conventional building materials will not adversely affect the fire risk to occupants and firefighters or the structural performance of buildings. However, fire safety of PV installations in buildings is, generally, not well covered in current building regulations and standards; the fire safety requirements are treated on a national level in varying degrees of detail [42]. Reaction to fire tests for BIPV façades, where the entire system is evaluated in its real-world configuration, are specified in building regulations across various countries. However, it should be noted that the majority of currently available BIPV products only provide compliance with IEC standards and may not have undergone comprehensive system-scale fire testing.

The international standard IEC 61730-2 includes test procedures to assess potential fire hazards due to the operation of a PV module or failure of its components. Fire hazard safety requirements for BIPV modules comprise aspects such as safety against ignitability, but also the endurance of the module to overload reverse currents caused by hotspots and high temperatures in their bypass diodes. Most of these tests are based on the standard IEC 61215-2 procedures or ISO 11925-2 (see Table 2.3). However, fire tests that affect BIPV and BAPV are regulated by local codes and typically assess the module's ability to resist fire from external sources.

Table 2.3 Fire hazard testing procedures included in IEC 61730-2.

IEC 61730 test	Title	Reference standard
MST 22	Hotspot endurance test	IEC 61215-2
MST 23a	Fire test	National/Local codes
MST 24	Ignitability test	ISO 11925-2
MST 25	Bypass diode thermal test	IEC 61215-2
MST 26	Reverse current overload test	-

It can be argued that the rapidly developing approaches to integrate PV modules into the built environment have not yet been completely absorbed into the current standards and regulations based on traditional building materials and designs. The regulations and standards distinguish between BIPV installations, where the PV modules are considered as building materials, and BAPV installations, where the PV modules are considered to be an external attachment to the building and not part of the construction. The available research on PV fire safety summarised in this section clearly illustrates that the fire safety of a structure is a result of the total system performance, where all the different materials and components interact and determine the consequences of a fire. Mitigating measures and well-engineered and documented solutions should be used when installing PV installations in the built environment to avoid excessive consequences of fires and maintain an acceptable overall fire risk.

The fire hazards represented by PV installations on buildings can be divided into four categories that will be discussed in the following sections: (1) ignition due to the operation of a PV module or failure of its components; (2) changed fire dynamics of the building skin caused by cavities behind the PV modules; (3) obstruction of firefighting and evacuation due to risk of electric shock from energised PV installations; and (4) resistance of BIPV modules to fire.

2.6.9.1 Ignition due to the operation of a PV module or failure of its components

Ignition can be caused by an electrical fault in the PV installation, as may also occur in other electrical installations. Different types of electrical faults that can cause fires in PV installations can be classified into ground faults, line-to-line faults, and arc faults [43]. These faults were further explored by Mohd Nizam Ong et al. in a fault tree analysis, where the connectors, insulators, inverters, and modules were the most common PV components to cause fires [44]. Most of the materials used in PV modules, such as glass, aluminium, and silicon, are not combustible, but the modules also contain some combustible polymer materials. The reaction-to-fire properties of PV modules have been studied in the cone calorimeter by several authors [45–49], where the total heat released from the combustible materials was typically in the range of 30–80 kW/m^2. This corresponds to a layer of approximately 1 mm of ethylene vinyl acetate (EVA), as is used in many PV modules. Even though the amount of combustible material in the modules is limited, the fire may start and propagate further to other materials in the building. PV modules have also been tested in different reaction-to-fire tests. Cancelliere et al. [50] combined the CLC/TR 50670 test method with test specimens cut from PV modules and used the SBI test setup described in EN 13823 as a calorimeter to assess the reaction-to-fire properties of combinations of PV modules and roofing surfaces. Boddaert et al. [51] tested BIPV modules according to EN 13823 and CEN/TS 1187 with electrically active and passive PV modules. Preliminary results showed an insignificant change in reaction-to-fire properties when the modules were tested under normal electrical operating conditions. Electric arcing was observed after the prescribed test time.

2.6.9.2 Changed fire dynamics of the building skin caused by cavities behind the PV modules

After the initial ignition, the fire dynamics of the construction plays an important role in how the smoke and flames can spread across the PV installation, into the building, and to neighbouring buildings. The flame spread on flat roofs with PV modules was experimentally studied by Kristensen et al. [52], who found that PV modules accelerated the flame propagation across the roof surface in different degrees, depending on the distance between the roof and the PV module. Also, the cavity below the PV modules for sloped roofs can promote fire spread upwards, due to the buoyant flow of flames and hot gases, and downwards, due to burning melted material [53]. The same principle also applies to vertical façades, where the ventilated cavity behind PV modules can promote fire and smoke propagation across the façade and inside the building envelope [54,55].

2.6.9.3 Obstruction of firefighting and evacuation due to risk of electric shock from energised PV installations

When there is a fire in a building with PV installations, the firefighters need to consider the risks associated with electric shock from the energised parts of the installation. PV may also affect evacuation, e.g., if the evacuation pathway is out through a PV façade. As the photovoltaic cells continue to supply voltage if they are exposed to light, and the fire may have damaged the electric insulation, hazardous electric voltages may occur at different exposed metal components. Firefighters contacting the high voltages generated by the PV installation would have some electric current flowing through their bodies. A current above 2 mA causes a startle reaction that can lead to following injuries. Higher current can cause loss of muscle control or cardiac arrest [56]. Getting access to fires in constructions behind PV modules can be challenging and may obstruct or delay the firefighters' efforts. As firefighting is a time-critical activity, the injury and damage from the fires can increase in such cases.

2.6.9.4 Fire resistance of BIPV modules

The fire resistance of two different BIPV modules with a 6 mm two-layer tempered glass construction was tested by Huang et al. [57], who found that they burned through in approximately 12 minutes when exposed to fire, according to the standard fire curve in ISO 834. Similar results were also reported by Ishii [58]. In constructions where the BIPV modules are used as a fire-separating construction, the fire resistance with respect to the insulation and integrity should be assessed. If fire integrity of more than 12 minutes is required, further development will be necessary to prevent the glass from breaking and falling out of the module frame.

2.7 Aesthetics and architecture

BIPV offers architects an array of design opportunities to create visually striking and innovative structures. Solar panels can be seamlessly integrated into various building elements such as roofs, façades, windows, and canopies [59], enhancing the architectural character, as described in detail in the following chapters of the book. The interplay of light and shadow, as well as the dynamic patterns created by BIPV installations, can contribute to the uniqueness of a building's visual identity. By embracing BIPV as an artistic medium, architects can craft structures that not only generate electricity but also engage the senses and provoke thought.

One of the foremost considerations when integrating BIPV is to ensure harmony with the existing **architectural context.** Architectural styles, materials, and cultural influences must be taken into account to avoid aesthetic discordance. Architects should collaborate closely with BIPV engineers (such as façade planners) and manufacturers to select photovoltaic materials

Figure 2.6 Three stages during a large-scale fire test of a BIPV façade according to SP FIRE 105. Photo credits: RISE Fire Research.

and finishes that complement the building's design language. This collaborative approach ensures that BIPV installations enhance, rather than compromise, the overall aesthetic appeal of the structure.

While aesthetics is important, **the functional aspects** of BIPV integration cannot be overlooked. Architects must be aware of the impact of orientation, tilt, and layout of solar panels and the influence of shadow on the modules to optimise electricity generation[1] without sacrificing the visual appeal. This requires a keen understanding of the building's solar exposition and the local climate. BIPV systems should be seamlessly integrated into the building envelope to create a cohesive and efficient energy-converting solution. Architects have the creative flexibility to play with transparency and opacity when designing BIPV installations. Transparent solar panels can be employed in windows, skylights, or canopies, allowing natural light to filter through while generating electricity. On the other hand, opaque panels integrated into façades or roofs may offer a more consistent aesthetic while still contributing to electricity generation. The strategic combination of these approaches allows architects to balance aesthetics, daylighting, and energy efficiency.

Key factors that a BIPV planner should consider when integrating BIPV in a way that aligns with aesthetic goals and architecture are:

Architectural Compatibility: The BIPV design should harmonise with the architectural style and context of the building. The planner should work closely with architects to ensure that the solar elements complement the overall design and do not clash with the building's aesthetic language.

Material Selection: The choice of photovoltaic materials and finishes plays a crucial role in achieving the desired aesthetic. BIPV panels come in various colours, textures, and transparency levels. The planner should collaborate with architects and manufacturers to select materials that match or enhance the building's visual appeal.

Placement and Layout: Strategic placement of BIPV elements is essential. Panels can be integrated into roofs, façades, windows, or canopies. The planner should consider the building's orientation, solar exposition, and architectural features to determine the optimal layout for electricity generation while maintaining a pleasing visual balance. BIPV modules can be installed to be planar within the building surfaces or with different orientations to create shadows and patterns on the roof or on the façade.

Scale and Proportion: The size and proportion of BIPV elements should be proportionate to the building's scale. Oversized or undersized panels can disrupt the aesthetic harmony. Careful consideration should be given to the spacing and arrangement of panels to create a visually cohesive appearance.

Shadow Patterns and Light Play: BIPV installations can create captivating shadow patterns and interact with natural light in unique ways. The planner should explore how the changing position of the sun throughout the day affects the visual dynamics of the BIPV elements and how these interactions contribute to the building's aesthetic allure.

Transparency and Opacity: Depending on the design intent, the planner can choose between transparent, semi-transparent, and opaque BIPV materials. Transparent panels allow light to pass through, creating interesting lighting effects, whereas opaque panels might provide a more coherent visual surface.

Colour Palette: The colour of BIPV panels can be coordinated with the building's colour palette. Alternatively, the planner can explore contrasting colours to make the BIPV elements stand out as design features, if in line with the architectural concept.

Finally, when integrated into the building element, two other aspects should be considered both by the planner and by the building owner or the client: **Community and Public Perception**, and **Education and Awareness.** BIPV installations are seen not only by building occupants but also by the public. The planner should consider how the BIPV design contributes to the community's perception of the building and its role in sustainability. Educating stakeholders about the aesthetic benefits of BIPV can foster support and appreciation for the design. The planner can highlight how BIPV enhances the building's visual identity and aligns with broader sustainability goals.

By meticulously considering all of these factors, a planner can ensure that the integration of BIPV enhances the building's aesthetic appeal while also contributing to its energy efficiency and sustainability objectives.

2.7.1 Architectural objectives

The overall goal for architects is to combine aesthetic qualities with functional (usability) and technical aspects into one design. From an architectural viewpoint, the integration of PV has different sub-objectives:

- Combining technical functionalities with aesthetics
- Improving the usability (indoor qualities, daylighting, indoor shading patterns, indoor climate, user acceptance)

- Achieving envisioned proportions of the envelope (façades and roof) or of the shape of the building
- Visual integration in the 'concept of the design'

The aesthetic qualities of a building are linked to the proportions of the façades or the proportions of the form of a building as a whole. Figure 2.7 shows three different principles of integration of BIPV in architecture followed by three references illustrating these approaches. A BIPV system in the overall design can be:

- Recognizable in the layout of the building envelope
- Part of the envelope as a whole
- Defining the building shape to optimise energy performance

The notion of a "design concept" is closely intertwined with the architectural philosophy of the designer, reflecting personal perspectives influenced by cultural factors. For instance, some architects believe that a building should visibly convey its electricity-generating capability, while others advocate for the seamless integration of renewable energy systems into the building's architectural aesthetic. These varying viewpoints naturally result in distinctly different architectural expressions and, consequently, a demand for diverse qualities of BIPV building products.

The energy performance of the building is expressed in the architecture of the building. Solar-shading devices can be prominent features of the building, displaying how electricity is generated (Figure 2.8).

Figure 2.7 SNFCC cultural complex by Renzo Piano Building Workshop Architects (left), Omicron headquarters by Dietrich Untertrifaller Architects (middle), and Terra Sustainability Pavilion by Grimshaw Architects (right). The SNFCC cultural complex is in Athens, Greece, the Omicron campus is in Vorarlberg, Austria and the Terra Sustainability Pavilion is in Dubai, UAE. Photo credits: Yiorgis Yerolymbos (left) and SUNOVATION (middle and right).

Figure 2.8 Tecnalia headquarters in San Sebastian (Spain) designed by Tecnalia in cooperation with Tjerk Reijenga – BEAR iD (left) and a garage in Gothenburg (Sweden) designed by Liljewall arkitekter (right). Photo credits: Vega de Seoane, Tecnalia (left) and Anna Svenson, Soltech Energy Sweden (right).

2.7.2 General aesthetic requirements

The aesthetic qualities of a building are linked to the proportions of its façades or its overall form. Aesthetic aspects that must be considered include:

Sizes and shapes

The range of sizes of BIPV building products is generally limited. A larger range of sizes would create more possibilities for fitting the panels into the layout of a building's envelope. The same goes for shapes. The availability of different widths and heights of modules enlarges the flexibility in design.

Jointing

The arrangement of profiles between the cells (the jointing) is an important visual quality of the layout of façades. In first-generation PV products, the jointing was not very well designed. Not only were products added to the building skin, but the pattern of joints also interfered with those of the other cladding systems.

The range of sizes and shapes can be based on standard sizes of frequently used building products or chosen based on aesthetic considerations.

Fixings

The elements used to fix the modules to the substructure are very important for the overall aesthetics of the system, although they may seem like a minor detail. Almost-invisible mounting solutions generally lead to more visual acceptance.

Combination with adjacent building products

BIPV does not yet combine well with other building products, not only in terms of available sizes but also because of the

Figure 2.9 Night view of the BIPV curtain wall of Pierre Arnaud artwork gallery in Lens, Switzerland, showing coloured LED projections. Designed by: architect Jean-Pierre Emery. Credits: Ertex Solar.

limitations of the substructure. First-generation BIPV systems were not designed to fit the adjacent cladding material of the façade or roof, and because of this, the systems did not blend in.

Detailing of edges and rims

The perimeter of a façade/roof is an important detail. Building products need to have proper detailing at the edges of façades and roofs. Façade systems using frames have to be precisely detailed to avoid the prominent appearance of the frames dominating the overall appearance of the design.

Light transmission (transparency)

Transparency of the whole façade can be enhanced by using BIPV products. Daylighting, glare, visual contact with the outside, and enclosure toward the outside are some of the functionalities achieved by transparent façades. (See chapter 2.4).

Transparency in BIPV modules can be achieved by the space between the cells or by the cell material itself (see Figure 2.8). The first option is the most usual. In crystalline silicon modules, it produces a typical light/dark pattern on walls and floors. Due to the high contrast, the product is less suitable for a work environment (desk work). Moreover, semi-transparent BIPV modules can be combined with LED to create coloured light façades at night (see Figure 2.9), illuminating the panel in different tones.

In thin-film modules, the light transmission is more homogeneous, which makes the product more suitable for a work environment. Further insights and approaches can be found in the work carried out in the years 2000–2012 by the IEA PVPS Task 7 and IEA SHC Task 41 expert groups, while the most recent advancements in BIPV are detailed within IEA PVPS Task 15.

Note

1 In the past (10–20 years ago), the efficiency was more important due to the high cost of PV. With today's cost, more freedom for the design is possible, and an efficiency loss of 10–15% or even 30% for façade integration can be accepted, as this allows more and better integration in the design.

References

[1] K. Berger, et al., "Analysis of requirements, specifications and regulation of BIPV. Report IEA-PVPS T15-08: 2019," *International Energy Agency – PVPS*, 2019 [Online]. Available: https://iea-pvps.org/key-topics/analysis-of-requirements-specifications-regulation-of-bipv/.

[2] V. Shabunko and P. Kovács, eds., "Fire safety of BIPV: International mapping of accredited and R&D facilities in the context of codes and standards, Report IEA-PVPS T15-15: 2023," 2023 [Online]. Available: www.iea-pvps.org.

[3] G. Eder, et al., "Coloured BIPV Market, research and development IEA PVPS Task 15, Report IEA-PVPS T15-07: 2019," 2019 [Online]. Available: http://iea-pvps.org/index.php?id=task15.

[4] N. Jakica, R. J. Yang, and J. Eisenlohr, eds., "BIPV design and performance modelling: Tools and methods, Report IEA-PVPS T15-09:2019," 2019 [Online]. Available: https://iea-pvps.org/wp-content/uploads/2020/01/IEA-PVPS_15_R09_BIPV_Design_Tools_report.pdf.

[5] T. E. Kuhn, C. Erban, M. Heinrich, J. Eisenlohr, F. Ensslen, and D. H. Neuhaus, "Review of technological design options for building integrated photovoltaics (BIPV)," *Energy Build.*, vol. 231, p. 110381, 2021, doi:10.1016/j.enbuild.2020.110381.

[6] B. Blasi, T. Kroyer, T. Kuhn, and O. Hohn, "The MorphoColor concept for colored photovoltaic modules," *IEEE J. Photovoltaics*, vol. 11, no. 5, pp. 1305–1311, Sep. 2021, doi:10.1109/JPHOTOV.2021.3090158.

[7] W. M. P. U. Wijeratne, R. J. Yang, E. Too, and R. Wakefield, "Design and development of distributed solar PV systems: Do the current tools work?," *Sustain. Cities Soc.*, vol. 45, pp. 553–578, Feb. 2019, doi:10.1016/j.scs.2018.11.035.

[8] ASHRAE, *ASHRAE Handbook – HVAC Applications*. Atlanta, GA: American Society of Heating, Refrigerating and Air-Conditioning Engineers, Inc., 2019 [Online]. Available: www.ashrae.org/technical-resources/ashrae-handbook/description-2019-ashrae-handbook-hvac-applications.

[9] N. Martín and J. M. Ruiz, "Annual angular reflection losses in PV modules," *Prog. Photovoltaics Res. Appl.*, vol. 13, no. 1, pp. 75–84, 2005, doi:10.1002/pip.585.

[10] N. M. Kumar, M. Samykano, and A. Karthick, "Energy loss analysis of a large scale BIPV system for university buildings in tropical weather conditions: A partial and cumulative performance ratio approach," *Case Stud. Therm. Eng.*, vol. 25, no. June 2020, p. 100916, 2021, doi:10.1016/j.csite.2021.100916.

[11] N. M. Kumar, K. Sudhakar, and M. Samykano, "Performance comparison of BAPV and BIPV systems with c-Si, CIS and CdTe photovoltaic technologies under tropical weather conditions," *Case Stud. Therm. Eng.*, vol. 13, no. December 2018, p. 100374, 2019, doi:10.1016/j.csite.2018.100374.

[12] C. Deline, "Partially shaded operation of a grid-tied PV system," in *2009 34th IEEE Photovoltaic Specialists*

Conference (PVSC), Jun. 2009, pp. 001268–001273. doi:10.1109/PVSC.2009.5411246.

[13] G. Lin, S. Bimenyimana, M. Tseng, C. Wang, Y. Liu, and L. Li, "Photovoltaic modules selection from shading effects on different materials," *Symmetry, vol. 12, no. 12, p. 2082,* 2020.

[14] G. de Oliveira e Silva and P. Hendrick, "Lead – acid batteries coupled with photovoltaics for increased electricity self-sufficiency in households," *Appl. Energy*, vol. 178, no. 2016, pp. 856–867, 2016, doi:10.1016/j.apenergy.2016.06.003.

[15] D. Masa-Bote, et al., "Improving photovoltaics grid integration through short time forecasting and self-consumption," *Appl. Energy*, vol. 125, pp. 103–113, 2014, doi:10.1016/j.apenergy.2014.03.045.

[16] F. M. Vieira, P. S. Moura, and A. T. de Almeida, "Energy storage system for self-consumption of photovoltaic energy in residential zero energy buildings," *Renew. Energy*, vol. 103, pp. 308–320, 2017, doi:10.1016/j.renene.2016.11.048.

[17] N. Martín-Chivelet and D. Montero-Gómez, "Optimizing photovoltaic self-consumption in office buildings," *Energy Build.*, vol. 150, pp. 71–80, 2017, doi:10.1016/j.enbuild.2017.05.073.

[18] P. Corti, L. Capannolo, P. Bonomo, P. De Berardinis, and F. Frontini, "Comparative analysis of BIPV solutions to define energy and cost-effectiveness in a case study," *Energies*, vol. 13, no. 15, p. 3827, Jul. 2020, doi:10.3390/en13153827.

[19] H. Ishii, "Thermal performance (G-value and U-value) – evaluation of BIPV applied to glass Façade," 2017 [Online]. Available: www.eupvsec-proceedings.com/proceedings?paper=42114.

[20] D. Valencia-Caballero, "Thermal transmittance and solar factor in BIPV operating conditions," *Presented at the 17th Advanced Building Skins Conference*. Bern, 2022.

[21] P. Sánchez-Palencia, N. Martín-Chivelet, and F. Chenlo, "Modeling temperature and thermal transmittance of building integrated photovoltaic modules," *Sol. Energy*, vol. 184, pp. 153–161, May 2019, doi:10.1016/j.solener.2019.03.096.

[22] E. Barykina and A. Hammer, "Modeling of photovoltaic module temperature using Faiman model: Sensitivity analysis for different climates," *Sol. Energy*, vol. 146, pp. 401–416, Apr. 2017, doi:10.1016/J.SOLENER.2017.03.002.

[23] D. L. King, W. E. Boyson, and J. A. Kratochvill, "Photovoltaic array performance model. Report SAND2004-3535," Alburquerque, USA, 2004, doi:10.2172/919131.

[24] D. Faiman, "Assessing the outdoor operating temperature of photovoltaic modules," *Prog. Photovoltaics Res. Appl.*, vol. 16, no. 4, pp. 307–315, Jun. 2008, doi:10.1002/pip.813.

[25] R. G. Ross, "Interface design considerations for terrestrial solar cell modules," in *IEEE Photovoltaic Specialists Conference*, Twelfth IEEE PVSEC, 1976, pp. 801–806.

[26] M. Fuentes, "A simplified thermal model for Flat-Plate photovoltaic arrays. Report Number: SAND-85-0330 ON: DE87009386," 1987 [Online]. Available: www.osti.gov/biblio/6802914.

[27] N. Martín-Chivelet, J. Polo, C. Sanz-Saiz, L. T. Benítez, and M. Alonso-Abella, "Assessment of PV module temperature models for building-integrated photovoltaics (BIPV)," *Sustainability*, vol. 14, no. 1500, pp. 1–15, 2022, doi:10.3390/su14031500.

[28] B. Roy, A. Gagné, and V. Delisle, "Thermal assessment of the varennes library building-integrated photovoltaic rooftop system," in *Proc. 5th COBEE, Montreal*, 2023, pp. 1061–1069, doi:10.1007/978-981-19-9822-5_112.

[29] L. Gaisberger, Roland Valckenborg et al., "IEC 61853-matrix analysis of PVPS Task 15 BIPV round-robin for more than one year at seven test sites over the world," in *37th European Photovoltaic Solar Energy Conference and Exhibition*, EU PVSEC, 2020, pp. 1719–1725, https://www.photovoltaic-conference.com; https://doi.org/10.4229/EUPVSEC20202020-6DO.12.1.

[30] N. Martín-Chivelet et al., "Building-Integrated Photovoltaic (BIPV) products and systems: A review of energy-related behavior," *Energy Build.*, vol. 262, May 2022, doi:10.1016/j.enbuild.2022.111998.

[31] H. Ishii, M. Kondo, and S. Hagihara, "Evaluation of thermal properties for BIPV in Façade proposal of SHGC evaluation method for BIPV using ISO 19467," 2019 [Online]. Available: www.eupvsec-planner.com/presentations/c48702/evaluation_of_thermal_properties_for_bipv_in_facade_-_proposal_of_shgc_evaluation_method_for_bipv_using_iso_19467.htm.

[32] L. Olivieri, F. Frontini, C. Polo-López, D. Pahud, and E. Caamaño-Martín, "G-value indoor characterization of semi-transparent photovoltaic elements for building integration: New equipment and methodology," *Energy Build.*, vol. 101, pp. 84–94, 2015, doi:10.1016/j.enbuild.2015.04.056.

[33] Commission Internationale de l'Eclairage, *CIE 13.3:1995. Method of Measuring and Specifying Colour Rendering Properties of Light Sources*. CIE, 1995.

[34] T. Reijenga, et al., *Successful Building Integration of Photovoltaics – A Collection of International Projects*. International Energy Agency, 2020 [Online]. Available: https:https://iea-pvps.org/wp-content/uploads/2021/03/IEA-PVPS-Task-15-An-international-collection-of-BIPV-projects-compr.pdf.

[35] A. Fedorova, B. P. Jelle, B. D. Hrynyszyn, and S. Geving, "A testing methodology for quantification of wind-driven rain intrusion for building-integrated photovoltaic systems," *Build. Environ.*, vol. 199, no. February, 2021, zdoi:10.1016/j.buildenv.2021.107917.

[36] CENELEC, *PD CEN/TR 15601:2012 Hygrothermal Performance of Buildings Resistance to Wind-Driven Rain of Roof Coverings with Discontinuously Laid Small Elements Test Methods*, CENELEC, p. 28.

[37] *EN 50583–2:2016 Photovoltaics in Buildings – Part 2: BIPV Systems*. European Committee for Electrotechnical Standardization, 2016.

[38] N. Martin Chivelet and I. Fernández Solla, *La envolvente fotovoltaica en la arquitectura: criterios de diseño y aplicaciones*. Reverté, 2007 [Online]. Available: https://books.google.com/books?id=Gpwzi3zgoUMC&pgis=1.

[39] G. A. Kopp, "Wind loads on building components and cladding," in *Advanced Structural Wind Engineering*, Y. Tamura and A. Kareem, Eds. Springer, 2013. doi:10.1007/978-4-431-54337-4_7.

[40] S. Misara and A. Pornnimit, "Mechanical characteristics of BIPV modules under different load scenarios and encapsulations," 2011 [Online]. Available: http://proceedings.ises.org/paper/swc2011/swc2011-0176-Misara.pdf.

[41] T. Stathopoulos, I. Zisis, and E. Xypnitou, "Local and overall wind pressure and force coefficients for solar panels," *J. Wind Eng. Ind. Aerodyn.*, vol. 125, pp. 195–206, Feb. 2014, doi:10.1016/j.jweia.2013.12.007.

[42] R. Yang, et al., "Fire safety requirements for building integrated photovoltaics (BIPV): A cross-country comparison," *Renew. Sustain. Energy Rev.*, vol. 173, no. December 2022, p. 113112, 2023, doi:10.1016/j.rser.2022.113112.

[43] M. K. Alam, F. Khan, J. Johnson, and J. Flicker, "A comprehensive review of catastrophic faults in PV arrays: Types, detection, and mitigation techniques," *IEEE J. Photovoltaics*, vol. 5, no. 3, pp. 982–997, May 2015, doi:10.1109/JPHOTOV.2015.2397599.

[44] N. A. F. Mohd Nizam Ong, M. A. Sadiq, M. S. Md Said, G. Jomaas, M. Z. Mohd Tohir, and J. S. Kristensen, "Fault tree analysis of fires on rooftops with photovoltaic systems," *J. Build. Eng.*, vol. 46, p. 103752, Apr. 2022, doi:10.1016/j.jobe.2021.103752.

[45] C. L. Chow, S. S. Han, and X. M. Ni, "A study on fire behaviour of combustible components of two commonly used photovoltaic panels," *Fire Mater.*, vol. 41, no. 1, pp. 65–83, Jan. 2017, doi:10.1002/fam.2366.

[46] L. Yin, Y. Jiang, and R. Qiu, "Combustion behaviors of CIGS thin-film solar modules from cone calorimeter tests," *Materials (Basel)*, vol. 11, no. 8, p. 1353, Aug. 2018, doi:10.3390/ma11081353.

[47] H.-Y. Yang, X.-D. Zhou, L.-Z. Yang, and T.-L. Zhang, "Experimental studies on the flammability and fire hazards of photovoltaic modules," *Materials (Basel)*, vol. 8, no. 7, pp. 4210–4225, Jul. 2015, doi:10.3390/ma8074210.

[48] X. Ju, et al., "Experimental study on burning behaviors of photovoltaic panels with different coverings using a cone calorimeter," *J. Renew. Sustain. Energy*, vol. 9, no. 6, Nov. 2017, doi:10.1063/1.4990830.

[49] T. Ohuchi, N. Ishikawa, and Y. Kozawa, "Improvement of the fire-proofing and fire-resistance properties of PV modules for building's exterior walls," in *Conference Record of the Twenty-Eighth IEEE Photovoltaic Specialists Conference – 2000 (Cat. No.00CH37036)*, pp. 1533–1538. doi:10.1109/PVSC.2000.916187.

[50] P. Cancelliere, G. Manzini, G. Traina, and M. G. Cavriani, "PV modules on buildings – Outlines of PV roof samples fire rating assessment," *Fire Saf. J.*, vol. 120, p. 103139, Mar. 2021, doi:10.1016/j.firesaf.2020.103139.

[51] S. Boddaert, F. Parolini, P. Bonomo, and M. Awais, "Most demanding standard combination to support BIPV development and sustainability," *Fire Saf. J.*, 2022.

[52] J. S. Kristensen, F. B. M. Faudzi, and G. Jomaas, "Experimental study of flame spread underneath photovoltaic (PV) modules," *Fire Saf. J.*, vol. 120, p. 103027, Mar. 2021, doi:10.1016/j.firesaf.2020.103027.

[53] R. Stølen, J. S. Fjærestad, and R. F. Mikalsen, "EBOB – Solcelleinstallasjonar på bygg. Eksperimentell studie av brannspreiing i holrom bak solcellemodular på skrå takflater. Del 2: Teknisk rapport, RISE Fire Research, RISE Rapport 2022:83, 2022," 2022, doi:10.13140/RG.2.2.34287.92326.

[54] M. Aram, X. Zhang, D. Qi, and Y. Ko, "Scaling study of smoke spread from building integrated photovoltaic (BIPV) double skin façade fire for achieving sustainable buildings and cities," *Sustain. Cities Soc.*, vol. 97, p. 104648, Oct. 2023, doi:10.1016/j.scs.2023.104648.

[55] R. Stølen, T. Li, T. Wingdahl, and A. Steen-Hansen, "Large- and small-scale fire test of a building integrated photovoltaic (BIPV) façade system," *Fire Saf. J.*, vol. 144, p. 104083, Mar. 2024. doi:10.1016/j.firesaf.2023.104083

[56] R. Backstrom and D. Dini, "Firefighter safety and photovoltaic installations research project," *Reliab. Photovolt. Cells, Modul. Components, Syst. V*, vol. 8472, p. 84720K, 2012, doi:10.1117/12.965313

[57] Y.-C. Huang, S.-K. Lee, C.-C. Chan, and S.-J. Wang, "Full-scale evaluation of fire-resistant building integrated photovoltaic systems with different installation positions of junction

boxes," *Indoor Built Environ.*, vol. 27, no. 9, pp. 1259–1271, Nov. 2018, doi:10.1177/1420326X17713256.

[58] H. Ishii, "Experimental study on BIPV module during power generation exposed to fire," 2018, https://userarea.eupvsec.org/proceedings/35th-EU-PVSEC-2018/6BV.1.44/.

[59] P. Bonomo, et al., "Categorization of BIPV applications. IEA-PVPS T15–12:2021," 2021 [Online]. Available: https://iea-pvps.org/wp-content/uploads/2021/09/IEA-PVPS-T15-12_2021_BIPV-categorization_report.pdf.

[60] A. Gok, E. Ozkalay, G. Friesen and F. Frontini, "The Influence of Operating Temperature on the Performance of BIPV Modules," in IEEE Journal of Photovoltaics, vol. 10, no. 5, pp. 1371-1378, Sept. 2020, doi: 10.1109/JPHOTOV.2020.3001181

[61] E. Cadoni, D. Forni, M. Dotta, G. Bellenda, and M. Caccivio, "Advanced characterisation of photovoltaics for hail resistance," Mater. Lett., vol. 354, no. June 2023, 2024, doi: 10.1016/j.matlet.2023.135371.

CHAPTER 3
BIPV products

Francesco Frontini, Pierluigi Bonomo, Nuria Martín Chivelet, Astrid Schneider, and Costa Kapsis

3.1 Introduction

This chapter describes the BIPV products from their components to their final layout as part of the building envelope construction systems. The BIPV module technologies mainly differ in the PV cell type and the front and back cover materials, but also important for the BIPV design may be the additional layers, such as the encapsulant or the coatings on the front and backsheet, and the variations in the density and distribution of the PV cells, which lead to different looks and a variety of transparency levels. The manufacturing techniques also allow variations in the module characteristics to fulfill the construction requirements. Currently, more than 95% of BIPV products on the market are based on crystalline silicon glass laminates, which allow a large variety of customisation levels to accomplish various architectural applications. The chapter shows the BIPV product design possibilities, challenges, and development trends for their integration into roofs, façades, and shading devices.

3.2 Components and assembly of a BIPV module

3.2.1 Photovoltaic cell types

The standard "IEC 63092 Photovoltaics in buildings – Part 1: Requirements for building-integrated photovoltaic modules" defines a BIPV module as follows:

> *A BIPV product or BIPV module is the smallest (electrically and mechanically) non-divisible photovoltaic unit in a BIPV system which retains building-related functionality. At the same time, it represents a functional unit of the building envelope satisfying the primary functions of the construction: if the BIPV product is dismounted, it would have to be replaced by an appropriate construction product.*

A BIPV module is a photovoltaic (PV) module and a construction product at the same time, mainly designed to be a multifunctional component of the building skin.

PV modules generate renewable electricity by directly converting solar radiation into direct current (DC) using semiconductor materials. PV modules are made of PV cells, which represent the principal elements responsible for the energy conversion in a BIPV product. They can be classified according to the cell or film technology, each one having different solar energy conversion efficiencies, design and appearance.

3.2.1.1 Crystalline silicon solar cells

Crystalline silicon PV cells (c-Si) have dominated the market for years. They are based on crystalline silicon wafers. Depending on the type of crystallisation, these wafers and the cells manufactured from them can be multicrystalline or monocrystalline. Multicrystalline silicon (mc-Si, also called polysilicon) wafers show crystals directed to different angles with a structure comparable to granite stones. Oppositely, monocrystalline silicon wafers show homogeneous surfaces because they come from the cut of large cylindrical single crystals. During cell manufacturing, the crystalline wafers (mono or multi) run through chemical and physical processes. The last step is the application of an anti-reflective coating, which turns the naturally occurred metallic grey-coloured surface of a silicon cell into the typical blue-to-black common appearance of c-Si PV cells.

Today, monocrystalline silicon PV cells, with conversion efficiencies exceeding 24% (and modules with over 20% efficiency), dominate

DOI: 10.1201/9781003432241-3

the photovoltaic market. This trend is expected to continue in the coming decade [1]. Whereas conventional solar cells are designed to harness solar energy from only one side, the current trend for utility-sized PV applications is rapidly shifting towards the use of bifacial PV cells (and modules), which are capable of harnessing solar energy from both sides, front and rear. In 2023, about 50% of produced PV cells were bifacial[12]. Typically, their rear side is about 20% less efficient than their front side. This trend brought a change as well in the solar module construction: while earlier standard modules had a white foil as a backsheet, nowadays, bifacial standard photovoltaic modules are mostly framed glass-glass-laminates to allow the entry of sunlight from both sides of the module.

3.2.1.2 Thin-film solar cells

In 2021, approximately 5% of the total PV modules in the market utilised thin-film technologies [13]. Unlike crystalline wafer-based modules, thin-film PV modules are composed of layers of transparent and opaque conductive PV materials, typically coated or sputtered onto a substrate. The manufacturing process for thin-film PV is similar to low-emissivity (low-e) hard-coated window glass. Thin-film PV cell materials are applied to a substrate – mostly transparent conductive oxide (TCO) glass – in several layers, with the coating order being from the front glass ("superstrate") or the back cover ("substrate"). Beyond TCO, stainless steel or aluminium can also be used as a substrate.

Thin-film PV can be made from different semiconductor materials such as amorphous silicon (a-Si), copper indium gallium selenide/sulfide (CIGS), and cadmium telluride (CdTe). While a-Si PV cell efficiencies are significantly lower than crystalline silicon, CIGS and CdTe efficiencies are already comparable to multi-crystalline silicon ones. In the early days of PV, amorphous silicon was the first and most widespread thin-film material. Its appearance, versatility regarding its properties, and lower manufacturing costs made this technology attractive. However, its low efficiency (6–8%), when compared to other commercialised technologies, has become a market barrier.

Typically, thin-film PV is opaque. For the film to become semi-transparent, layers of the PV material are removed by laser etching, widening the thin lines between cells. This way, the transparency is usually set between 10–20%. Although this process leads to a decrease in module efficiency, their visual appearance of homogeneous greyish semi-transparent elements makes thin-film technologies an appealing product for skylights and façade applications. Organic photovoltaic (OPV) is expected to also contribute to thin-film semitransparent BIPV solutions.

3.2.1.3 Heterojunction and multijunction PV cells

Most commercial PV cells are homojunction, meaning that they are made of just one semiconductor to form the cell junction that creates voltage. Alternatively, heterojunction PV cells include two semiconductor materials to form the cell junction. The silicon heterojunction (SHJ) technology combines the advantage of monocrystalline silicon and thin-film materials, which are coated on top. Typically, these thin-film layers are composed of amorphous silicon or perovskite. SHJ PV cells can achieve efficiencies above 24%. It's important to note that SHJ cells are classified as a type of crystalline photovoltaic cells.

A multijunction solar cell (MJSC) consists of several individual cells (sub-cells) stacked together and connected in series to obtain higher performance by combining different solar radiation wavelength sensitivities (spectral responses). An MJSC made of two cells is commonly named a tandem cell. Mostly, MJSCs are used in PV concentration systems.

3.2.2 PV modules manufacturing

Today, crystalline silicon PV modules have the largest market share, accounting for over 95%. Consequently, this section provides a detailed description of this technology. It is important to note, however, that thin-film modules share many common features with crystalline silicon ones.

Crystalline PV module manufacturing starts with the electrical connection of several PV cells in series, making up a "cell string." In a string, the same current goes through all of the cells (there is no current intensity addition), and cells add their voltages to sum the final voltage of the module. Usually, PV modules are made of more than one string to also increase the output current (several strings in parallel add their currents). All of the strings are connected at the junction box of the PV module, which also contains bypass diodes. These electronic devices allow PV current to "bypass" shaded cell strings or even the whole module. The issue with shaded solar cells is twofold: not only do they cease to produce electricity, but they also create hotspots. Figure 3.1 shows the main steps from the PV cell to the module, to the BIPV system.

To set up a BIPV system, the solar modules are connected in series to form "module strings." The module strings are then connected either in series or in parallel at the PV array combiner box. The BIPV system may also contain overcurrent protection and disconnection devices. Its output is the DC BIPV array (or BIPV generator) output, which can further be connected to a power conditioner equipment such as an inverter. Inverters transform direct current (DC) into alternating current (AC). More information about inverters can be found in Chapters 2 and 4.

Thin-film module manufacturing technologies differ from c-Si ones. In thin-film modules, the PV material layers are created by different techniques, such as printing, sputtering, or chemical

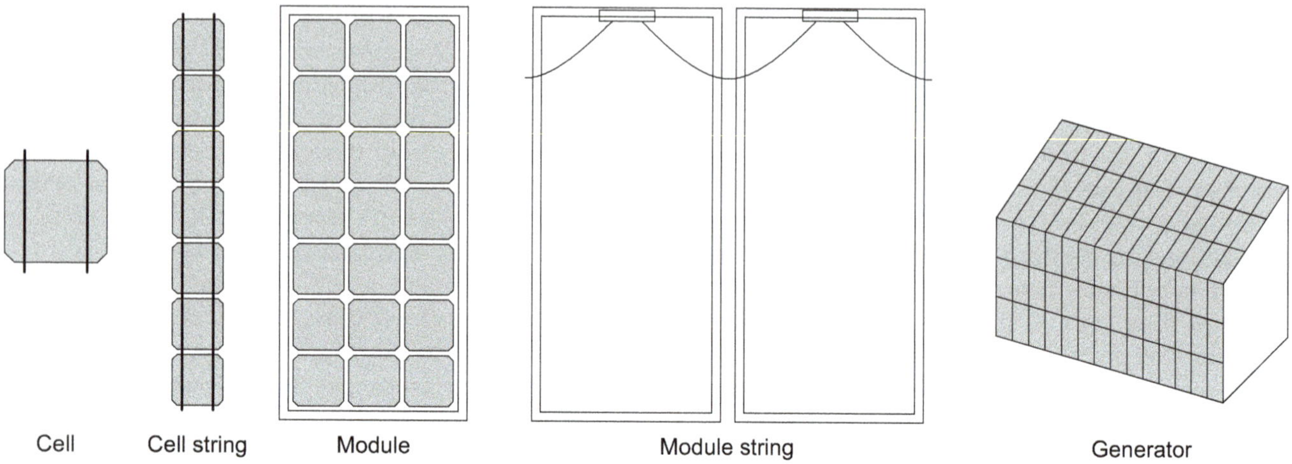

Figure 3.1 From PV cells to BIPV generator.

vapour deposition. Then, by laser ablation, PV cells are formed and interconnected, resulting in the appropriate module voltage and current values.

The common manufacturing process of a conventional (mass produced) PV module consists of the following steps (Figure 3.2):

- *Preparation of the raw material*: The various raw materials used for the assembly process are prepared and controlled.
- *Stringer machine and layup for PV module stringing*: The PV cells are placed in a solar stringer that interconnects the cells in series by soldering a coated copper wire, called ribbon, on the bus bar of the cell. The result is a PV cell string.
- *Placement on the glass*: PV cell strings are automatically or manually placed on the glass previously prepared with the first layer of encapsulant material.
- *Buses welding*: The bus ribbons for interconnecting PV cell strings are welded. This phase can be automated or customised according to the desired application.
- *Pre-lamination preparation*: The second encapsulation layer and a foil of electrically insulating material called backsheet are applied. Then, the terminal ribbons are kept out to be connected in the junction box in a subsequent step. At this point of the manufacturing process, it is important to perform some electrical tests and possibly an electroluminescence test to verify that there are no short circuits or broken PV cells inside the module. At the pre-lamination stage, it is still possible to correct potential deficiencies.
- *Lamination*: The multi-layer sandwich is transformed into one single unit thanks to the polymerisation of the encapsulating material. Laminators work at high temperatures and vacuum levels. The resulting product is called PV laminate.
- *Junction box application*: Junction boxes in BIPV modules have special designs and placement to facilitate architectural integration and facilitate the connections between modules.
- *Framing (optional)*: If requested or needed, a frame is applied around the PV module, directly with tape on the laminate or using silicone in the aluminium channel frame.
- *Final test*: After assembling, the PV module is inspected and tested.

3.2.3 BIPV modules manufacturing: standard sizing or customisation

There are different ways to design, construct, and manufacture BIPV modules. However, all of the methods share the need to encapsulate and protect the PV cells, mainly from oxygen and moisture, to avoid corrosion. Most available BIPV products use laminated glass to protect the solar cells on both sides and, at the same time, to give the modules the mechanical strength required to perform as construction products. Other cell types and architectural applications can lead to additional encapsulation layers, different module materials, and construction designs.

There are three main approaches to manufacturing BIPV modules:

1. **Mass production for standardised products:** The layering of BIPV products remains fixed, allowing only one module layout per production line. It involves a series of consecutive operations performed by automatic machines dedicated to optimising each production phase, transforming raw materials into finished products. The process typically comprises the eight or nine steps presented previously.
2. **Full customisation for custom designs:** BIPV modules can be produced in various sizes, shapes, and with various layers of materials, for example, with different glass types and colour layers. In this case, several steps in the manufacturing line are manual with machine support.

Figure 3.2 (from left to right and top to bottom): "Half-cut" crystalline silicon PV cells; PV cells prepared for stringing; preparation of the cell layout of the module; standard high-efficiency PV modules with backsheet foil (left), and with black backsheet foil for "all black" appearance (right). Photos taken at the Sonnenkraft GmbH facility in Austria (credits: Astrid Schneider, TU Wien).

The level of customisation impacts the final cost of the product.

3. **Mass customisation:** The mass customisation approach aims to minimise material and size variations while employing a semi-automatic assembly line capable of producing most of the solutions required in the market. Automated steps alternate with manual work.

Overall, the manufacturing and assembly of BIPV modules involve a variety of techniques and options that cater to different design requirements and market demands, considering factors such as functionality, aesthetics, and sustainability.

The following simile can be made between the effort and cost of these PV production methods and fashion: readymade mass-produced jeans in a big store are cheaper than tailor-made ones. However, specialised shops such as sport shops might offer specialised pants for reasonable prices. This might help to understand the differences between PV and BIPV markets.

Large PV manufacturers – with about 95% located in China – produce "standard PV modules" produced and optimised for utility-scale PV plants or large rooftop applications. In PV plants, about 60% of modules are currently larger than 2.5 m^2 (e.g., PV modules with nominal power of around 700 Watts are up to 3.1 m^2), whereas for rooftop installations, module sizes are smaller than 2 m^2 because of manual installation limitations [3] and restrictions by building authorities such as the German Institute for Construction "DIBT-Deutsches Institut für Bautechnik."

In earlier times, it was easier than today to differentiate BIPV modules from PV modules with a quick look, because standard PV modules were always equipped with a frame and a white backsheet. Nowadays, due to the innovative bifacial PV cells, which can receive solar radiation and produce solar electricity from both sides, most advanced and highly efficient standard PV modules use glass as a backsheet to allow maximum exploitation of back-reflected solar radiation. As those modules are optimised for maximum electricity production with minimum use of materials, the frontsheet and backsheet glasses are extremely thin with 2 mm thickness. In this case, the structural strength of the module is normally provided by an aluminium frame. The producers deliver the modules with an installation manual, which specifies where and how to fix them to guarantee wind and snow load resistance.

3.2.4 Differences between standard PV and BIPV modules

In most building rooftop PV installations, standard PV modules are commonly used. This is primarily because these installations are categorised as building-attached photovoltaics (BAPV), where the modules are affixed to an existing building envelope without serving any other function than renewable electricity generation. The use of PV in buildings, including both BAPV and BIPV, is on the rise. In 2020, the SolarPower Europe Association reported that 44% of the global installed PV power was mounted on buildings, presenting a significant growth opportunity for BIPV.

Now, the key question is: what differentiates a BIPV module from an ordinary standard PV module? The answer is easy: a BIPV module is designed, constructed, and certified according to both electrotechnical and building construction requirements [3].

Most BIPV modules are based on glass-glass PV laminates. Additional attributes that differentiate BIPV from standard PV modules are as follows:

- Frameless: The module strength is provided by the laminate.
- Thicker frontsheet and backsheet glasses of both 3 mm, 4 mm, or even 6 mm thickness and more, depending on the project and construction requirements
- Wider cell-free borders to avoid cell shading by the construction profiles
- Certified as a laminated safety glass and as a construction component by public construction authorities to allow the installation in building skin (e.g., overhead installation or even point fixings)
- Designs do not always prioritise module efficiency but rather aesthetic requirements (e.g., transparency, colour, thermal insulation, see Chapter 2).
- Junction box located on the edge of the module to allow an invisible wiring during installation.

As discussed in Chapter 2, the BIPV product requirements derive from:

- Electric safety requirements, based on low voltage electrotechnical regulations
- Laminated safety glass standards for products containing at least one glass pane
- Building requirements, as introduced by construction product regulation and building codes (e.g., fire safety requirements for roofs and façades, heat retention, mechanical safety, etc.)

Moreover, depending on the BIPV application, other requirements can apply, such as mechanical resistance and stability, health and the environment, safety and accessibility in use, protection against noise, energy economy, and heat retention. An example is the rain tightness for BIPV roofing elements such as PV roof tiles.

The question about which standards must be fulfilled is strongly dependent on the use case and the function the PV modules shall fulfil, the building type and height and, especially, the local and national construction codes, regulations, and guidelines, which define the required qualifications and relevant certificates.

In the European Union, the reference standard for BIPV products and systems is EN 50583-1 and -2, respectively. In 2020, a similar international standard was also introduced:

IEC 63092-1 and -2, adopted by several countries across the globe.

3.2.5 Design options regarding colour, transparency, and reflectivity

BIPV modules tend to be installed on the "sunny side" of buildings, mostly highly visible, when it comes to façades, shading elements, or canopies.

When incorporating photovoltaics as building components, designers and architects seek various options to properly integrate them into the building's surfaces. An important question is how far the designer can influence the appearance of the solar modules regarding colour, shape, and reflectivity to fit the building's design qualities. To enrich BIPV products, from standardised manufacturing approaches to flexible and customised options, the industry has developed different solutions by modifying one or more of the typical layers of the module:

- Front cover (*frontsheet*)
- Encapsulant
- Crystalline solar cell strings
- Encapsulant
- Back cover (*backsheet*)

Each of the layers – and inner and outer surfaces of the front and back glasses – can be varied to change the module design and appearance. Photovoltaic cells can differ in colour, shape, dimensions, and placement, while other layers can be modified using different types of printings and coatings for front glasses, interlayers, or backsheet materials, leading to a variety of colours and transparency levels (see Figure 3.3). Several sophisticated solutions combine measures on different layers of the module to achieve the desired appearance.

3.2.5.1 Frontsheet

The frontsheet of a c-Si module or a thin-film module typically features a 2 mm to 6 mm glass cover with high solar transmittance, allowing maximum solar radiation to pass through and reach the photovoltaic layer. Known as low-iron (or solar) glass because of its low iron oxide content, the frontsheet is always heat-treated or tempered. To reduce the frontsheet reflections, an anti-reflective coating (ARC) can be implemented, featuring a macroscopic surface texture in the form of micro-pyramids or micro-domes that act as a "light trap." Using ARCs, the solar transmittance of the frontsheet glass can reach up to 95%, which compares favourably to approximately 92% for low-iron glass without ARC of the same thickness [4]. Note that ARC will increase soiling for BIPV installations that are not vertical (e.g., roofs, solar shading), due to surface texture and the hydrophilic nature of the ARCs currently used in the market.

In recent years, spectrally **selective coloured films and coatings** and **glass treatments**, such as **digital printing, etching, or frit**, have been applied on the glass frontsheet to provide a colour finish or pattern on the BIPV module. In these cases, module efficiency is sacrificed for aesthetics, due to shading of the PV layer, when compared to a clear frontsheet module [5]. When a colour or a pattern treatment is applied, the photovoltaic layer is fully or partially "camouflaged" behind the BIPV frontsheet. This opens new architectural possibilities that enhance building aesthetics and turn building surfaces into renewable energy generators. Note that other dated methods exist for producing coloured BIPV methods, some of which are discussed in the following sections.

Frontsheets for flexible photovoltaic modules can be made from transparent plastic materials, such as fiberglass-reinforced plastic (Figure 3.4). The same material can also be used as the backsheet. Other backsheet options used to manufacture

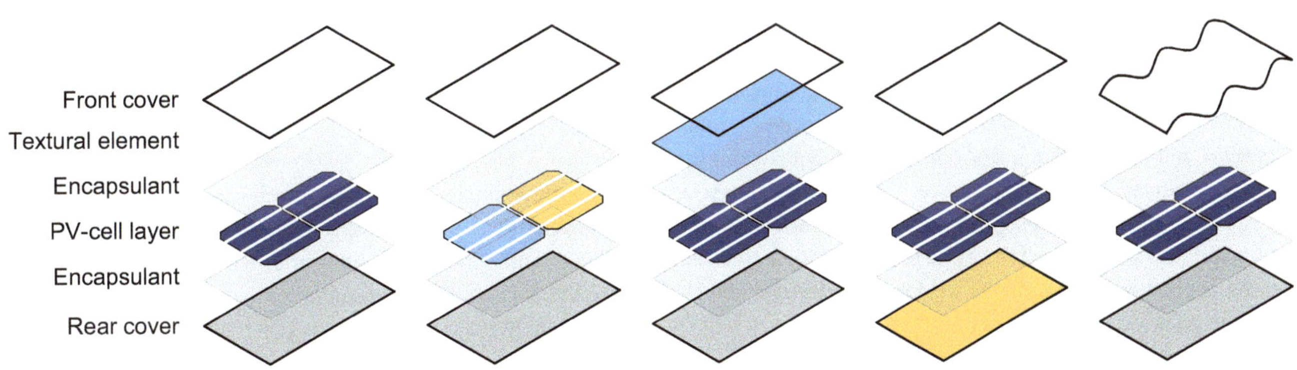

Figure 3.3 Schematic layer representation of a conventional PV laminate (left) and four alternative BIPV designs.

Figure 3.4 Lightweight flexible photovoltaic module with thin monocrystalline solar cells encapsulated between two fibre-reinforced plastic sheets; see the roof project in Figure 3.7. Photo: DAS Energy, Austria.

flexible thin-film PV modules or super-thin c-Si modules are flexible metal sheets or synthetic substrates.

3.2.5.2 Encapsulation

The encapsulation layer serves multiple crucial functions, including providing structural integrity between the frontsheet, PV layer, and backsheet, as well as protecting against the intrusion of humidity and oxygen into the PV laminate, as it would cause corrosion to the sensitive solar cells. Common encapsulation materials include ethylene vinyl acetate (EVA), primarily used in glass-laminate modules, and polyvinyl butyral (PVB) that is commonly used for glass-glass modules. Poly olefine elastomer encapsulants (POE) is a newer encapsulant material used in bifacial and thus glass-glass-laminates, as it is more resistant to chemical changes and degradation. Less commonly, silicone gel is used as an encapsulant. Some of the key attributes of encapsulation materials are to maximise solar transmission, achieve strong adhesion and cross-linking, and ensure long-term stability, especially under UV light exposure. To prevent moisture infiltration, which could lead to issues such as delamination and potential semiconductor device degradation, a sealant can be applied along the module's edges to effectively block moisture penetration.

Coloured encapsulants can be used to provide colour to semi-transparent thin-film technologies, but also to c-Si modules. The coloured encapsulants can be semi-transparent or opaque. Some only appear opaque in front of the solar cells but are indeed semi-transparent. Opaque encapsulants behind the cell layer are used to create, for example, a homogeneous black appearance of the overall solar module.

3.2.5.3 Photovoltaic layer

The photovoltaic layer consists of the solar cells and their interconnections. To obtain "full black" solar modules, it is necessary to hide the metal interconnectors between the cell strings. This can either be done by screen printing on the inner side of the front glass or by covering the metal connectors with small pieces of black foil. Another option is the use of so-called back contact silicon solar cells, which are connected behind the cells and do not show visible metallic grid and buses from the front side.

Crystalline silicon solar cells can be coloured by varying the thickness of the anti-reflective coating, from regular bright blue to gold, green, or a range of other colours. Nevertheless, the use of coloured cells results in a reduced module efficiency between 5% and 15%. The highest efficiency is attained with the optimised blue/black anti-reflective cell coating, which gives the well-known appearance to conventional PV modules.

Solar cell layout: different densities, transparencies, and patterns in PV cell distribution
In general, the lower the PV layer density and the higher the transparency, the lower the PV module efficiency. There are no transparent high-efficiency PV modules. Some novel near-transparent BIPV products use the frontsheet to concentrate sunlight onto the module edges where the PV layer is. So far, the efficiency of these novel products remains low (5% or less).

PV cell distribution density in BIPV modules can vary from maximum dense packing to lower cell densities, leading to 10%, 20%, 30%, or more daylight penetration. Furthermore, the solar heat gain coefficient of the BIPV will rise with its transparency. Customised solar cell patterns can be tailor-made (Figure 3.5). In this case, the cell strings have to be connected manually. Finally, the BIPV module shape can be varied by using different glass formats and geometries, such as circular, triangular, or free-form glasses.

3.2.5.4 Backsheet

Glass is the most common material used also as a backsheet in BIPV modules. Another common material is polyvinylidene difluoride (PVDF), a thermoplastic fluoropolymer-based foil. In the case of a PVDF, the backsheet tends to be opaque, and it is used in Si-based modules. In standard PV modules, PVDF is white to reduce the amount of solar irradiation absorbed by the backsheet and, thus, reduce the operating module temperatures.

Figure 3.5 BIPV skylight at the Schönbrunn Zoo, in Vienna, using customised solar cell patterns. Photo: Costa Kapsis.

For BIPV applications, PVDF can be coloured, creating from an all-black (e.g., black cells/black backsheet) to a plaid look (e.g., blue cells/red backsheet). The use of coloured opaque backsheets tends to marginally increase the operating module temperatures. Glass-foil BIPV products have less mechanical strength.

For opaque BIPV modules, coloured glass is often used as a backsheet to produce a coloured appearance. The colouring can be made by different means: by opaque enamelled screen printing on the outer surface of the glass or by coloured glass itself. Coloured and all-black designs are primarily used for rainscreen façades where no solar irradiation can be gained from the rear side. However, in the case of BIPV products for solar canopies and shading devices, that use bifacial solar cells to benefit from, e.g., ground albedo to increase annual electricity yield, the use of low-iron clear glass is preferred.

3.3 BIPV envelope solutions

3.3.1 BIPV roofing products

Roofs have the highest solar potential of buildings and are therefore not only the preferred place of installation for PV systems but also usually the most economical solution to generate solar electricity. Several BIPV systems have been developed for all kinds of roof construction types. As identified in the IEA-task 15 report "Categorisation of BIPV Applications" [7], the main roofing systems can be referred to as discontinuous roofing, continuous roofing, and skylights.

3.3.1.1 BIPV discontinuous roof

A pitched or sloped opaque roof is commonly covered with tiles or slates. This is a widespread roof construction method, referred to as a "discontinuous roof," due to the presence of small elements (tiles, slates, etc.), with the main function of water drainage. Pitched roofs are areas of the building envelope where PV deployment has been particularly successful. This success is attributed to various factors, such as high sun exposure or the ease of installing PV products with standard mounting systems on roofs, showing a consequently good economic payback. Many "in-roof" solutions have been developed in recent years, in which BIPV modules replace the traditional roof tiling layer. Solar tiles, metal sheets, shingles, slates, and prefabricated roofing modules that constructively replace traditional roof components are some of the possible applications. Although a solar roof is in the public imagination – a pitched simple roof of a single-family house with a few solar modules – more and more elaborated and innovative solutions are entering the market and are used in complex and ambitious designs. Some of the current main trends of innovation are related to:

- Aesthetic evolution of solar roof tiles due to different colours such as black, grey, red (terracotta), and different glass treatments, such as anti-reflective surfaces (see Figure 3.6) or structured glazing
- Availability of many different roofing tile typologies and sizes
- Customisation of solar tiling in complex roofs
- Availability of dummies without any electricity production (see Figure 3.6)

3.3.1.2 BIPV continuous roof

A planar or low-slope roof, often referred to as a "continuous roof," is distinguished by its continuous water-resistant layer, typically utilising membranes or metallic roofing. BIPV products have been used for these applications. BIPV continuous roof applications can be found in industrial

Figure 3.6 Left: Renovation featuring terracotta BIPV roof tiles on a historic farm in Ecuvillens, Canton of Fribourg, Switzerland; middle: solar shingles in a discontinuous roof; right: special roof parts for the Umwelt Arena in Spreitenbach, Switzerland. The modules are in an "all black" appearance with a black backsheet and covered interconnectors between the cell strings. Architect: Rene Schmid. Photos courtesy of Patrick Heinstein, CSEM (left) SUPSI; Flisom-Schweizer (middle); P. Bonomo (right).

halls, shopping centres, and airports, where flat roofs cannot carry heavy loads but offer large unshaded areas with high solar potential (Figure 3.7). The main products today are crystalline solar cells integrated into a laminate, where the front and backsheets are made from fibre-reinforced plastic that is flexible and can be used for flat or curved surface installations. Similar flexible solar modules are made with thin-film solar cells as well as encapsulated in plastic. This module type is then attached to either metal sheet roofing elements or roof membranes.

Additionally, there are innovations such as solar floors, which enable the installation of PV as a walkable roof, as illustrated in Figure 3.8.

Specific challenges are related with this application typology, such as anti-slippery treatments, user safety against falling and breakage, shading tolerance (if the roof is accessible), etc., so that components are specifically designed for the application field both in construction and electrical terms. Main market targets are multifamily or commercial buildings, and all buildings with available roof space that don't want to fully occupy the roof with solar modules but leave the possibility of using the roof.

3.3.1.3 BIPV skylights

Skylights or glass roofs are ideal places for the integration of PV modules, either as the main glazing for single-glazing roofs or as the outermost pane of an insulated glazing unit (Figure 3.9). The solar cells allow better control of important comfort parameters, such as solar gains and daylighting. The solar heat gain coefficient and the roof's insulation are important planning parameters. In both crystalline and thin-film technologies, the thermal/acoustic insulation and the transparency can be adjusted according to the indoor environmental requirements, depending on the climatic zone, the building typology, the internal functions, etc.

3.3.2 BIPV façades

Façades present a compelling option for PV integration, offering both advantages and disadvantages compared to roof installations. One notable drawback is that façades are more susceptible to shading, whether from surrounding buildings, self-shading, or trees. Furthermore, façades typically receive between 35% to 45% less irradiation than an optimally oriented roof would, depending on the geographic location. In regions with latitudes of 45 degrees or higher, BIPV façades demonstrate significant potential due to the consistently low sun angle throughout the year. Closer to the equator, the sun's path becomes more extreme, moving vertically over buildings and affecting the efficiency of façade-based PV installations. Another aspect to consider is that BIPV façades are intensively intertwined with the design of the building and its structural and design grid patterns, which are mostly determined by the height of the storeys and the width of the rooms. Typically, standard-size PV modules are usually not designed to meet the grid and design requirements of a building. For this reason, custom sized and designed BIPV modules are often chosen for façades, making them more expensive than conventional PV modules. Key reasons why façades are becoming an increasingly attractive option for BIPV applications can be summarised as follows:

Figure 3.7 Flat inclined roofs of the company "Trumpf" in Germany at the Rostock harbour. The crystalline solar cells are integrated into plastic laminates, incorporated in aluminium sheet roofing. The PV modules' weight is less than 4 kg per m². Picture: DAS Energy Ltd., Wiener Neustadt, Austria.

Figure 3.8 Walkable BIPV floor and balustrade realised with laminated safety glass and c-Si solar cells. Source: BIPVBOOST.

Cost-Competitiveness with Conventional Façade Systems: For many modern façade typologies, such as rainscreen, curtain walls, and double-skins, the costs of conventional systems (using materials like metal, fibre cement, ceramic, natural stone, or glass) are often on par with, or even exceed, those of BIPV systems. In this case, BIPV façades become cost-effective, as the marginal extra investment is quickly offset by energy savings.

Energy Generation Profile Aligned with Self-Consumption Needs: BIPV façades are particularly effective in terms of self-consumption. They can better align the energy generation profile with the building's load profile throughout different seasons and times of the day, unlike roof-based systems.

Some of the current main trends of innovation in façades are related with:

Cladding customisation: custom sized and designed BIPV modules are produced in a project-specific manner.

Camouflage BIPV elements: in the first "age of integration" of BIPV modules, a "showy" type of integration of the PV modules was preferred due to a mix of both: the wish to show the pretentious PV elements and a lack of preferable design options. Furthermore, it was generally the wish to maximise the solar gain. Today several technologies have become trendy: the BIPV cladding became "invisible" thanks to the use of coloured coatings and treated outer glass surfaces, allowing the full disappearance of the solar cells in the BIPV modules. Such a "camouflaged" and "designed" approach allows hiding the PV cells behind opaque and coloured patterns (Figure 3.10) [5].

Other trends of innovation focus on the future coupling of active PV claddings with sandwiched façade systems, including

Figure 3.9 BIPV skylight application using insulated glazing with lower cells density for daylight transmittance at the "house of choice" designed by White Architects in Stockholm.

Photo: Jesper Westblom.

Figure 3.10 Examples of coloured BIPV cladding for a façade where coloured glass is used to partially or fully cover the solar cells behind. Photo: courtesy of SUPSI.

thermal insulation layers, which are aimed at obtaining a totally dry installation, ensuring simplicity and efficiency of mounting for a unitised kit with thermal protection, fire prevention, and sound insulation [7].

3.3.2.1 Ventilated façades or rainscreen cladding systems

A ventilated façade or a rainscreen cladding system (also called "cold" façade") typically comprises a load-bearing building structure, insulation followed by an air gap, and an outer cladding system.

In a BIPV façade, the conventional cladding elements are substituted with PV modules (Figure 3.11). The market offers a multitude of construction models and technological solutions, each featuring various fixing options.

Typical market segments are multi-storey residential, commercial, and institutional buildings, both in new or renovation cases. The market offers a multitude of construction models and technological solutions, each featuring various joint types and fixing options.

3.3.2.2 Double-skin façades

A double-skin façade consists of two layers, usually made of glass, wherein air flows through the intermediate cavity. This space (which can vary from 20 cm to more than 1 meter) acts as a buffer zone against extreme temperatures, winds, and sounds, improving the building's thermal efficiency for both high and low temperatures. BIPV is applied similarly to a curtain wall even though the outer façade, in this case, does not require thermal insulation. Thus, it is often a glass laminate rather than an insulation glazing. A particular application of this technology is represented by bifacial solar cells. Bifacial façades take advantage of the recent development of bifacial PV cells, which can harvest sunlight from both sides (front and rear).

One of the current main challenges for PV elements integrated in high-rise buildings is fire safety. Ongoing research and normative developments are addressing aspects related to the possible reduction of some critical risk of electrical fire ignition, fire spread due to the combustible parts of PV modules and other parts, and regarding safety questions for both the maintenance and the rescue teams (Figure 3.12).

3.3.2.3 Curtain walls

A curtain wall is an external and continuous building skin fenestration system, totally or partially glazed, composed of panels supported by a substructure in which the outer components are non-structural. A curtain wall refers to its construction since the façade is hanging (just as a curtain) from the top perimeter of the building and is locally fixed to resist air and water infiltration. It is typically designed with extruded aluminium frames (but also steel, wood, etc.) filled with glass panes. The façade satisfies multiple requirements, such as a load-bearing function, acoustic and thermal insulation, light transmission, waterproofing, etc. In the case of a "warm façade," it divides, as a unitised skin layer, outdoor and indoor environments. It can be realised according to different construction systems such as stick-system, unitised curtain wall, Structural Sealant Glazing (SSG), and point-fixed or suspended façade. PV is typically part of the outer cladding layer, in the form of glass-glass elements, with both crystalline (Figure 3.13) or thin-film technologies and with various transparency degrees and visual appearance possibilities. Usually, the glass is an IGU (double or triple glazing) to ensure adequate thermal insulation. In the basic cases, it can be assimilated to a window. Thus, in windows, PV can be integrated into conventional PV glazing similarly to a curtain wall or also into some innovative applications.

3.3.3 BIPV shading

Photovoltaic cells and modules can be used as external photovoltaic shading systems to control light and reduce solar heat gain in curtain wall and double-skin façade solutions,

Figure 3.11 Rainscreen cladding with coloured modules in a multi-family house in Zurich, Switzerland, designed by Kämpfen Zinke + Partner AG. Photos: courtesy of I. Zanetti, SUPSI.

Figure 3.12 Double-skin façades featured in a new office building in Milan, Italy, and the renovated CSEM building in Neuchâtel, Switzerland. Photos: courtesy of P. Bonomo, SUPSI.

either as fixed elements or movable devices to adapt to dynamic solar conditions. Accurate simulation studies are crucial for designing these systems effectively, avoiding overshading, optimising solar harvesting, and ensuring desired thermal and visual comfort.

Large shading systems have been commonly used, as demonstrated in previous projects and studies such as [8–10], which analysed different south-facing shading solutions in office buildings. Movable BIPV shading systems (Figure 3.14) are also employed to control solar radiation and allow solar energy to enter the interior, based on cooling and heating demand. Various architectural shading devices that incorporate photovoltaic technologies have been invented, indicating the growing interest in integrating renewable energy generation into shading systems.

Lightweight photovoltaic materials such as CIGS offer flexible and curved shapes, making them suitable for dynamic shading systems. Although CIGS has lower energy conversion efficiency compared to traditional c-Si modules, it is advantageous for buildings with large, glazed areas or multi-storey structures that require shading systems and have sufficient transparent area. Integrating PV modules into dynamic shading systems enables the fine-tuning of different functions, such as generating electricity, balancing energy performance, and expressing architectural design while ensuring visual and daylight comfort.

Currently, several innovative concepts represent the frontier of dynamic active shading devices. One concept is external movable shading developed by the Polytechnic of Zurich ETHZ, which utilises a soft-pneumatic actuator to adapt blind positions based on user needs and desired comfort (Figure 3.14). Another concept, developed by SUPSI, involves integrating the photovoltaic blind into an insulating glazing unit, providing long service life and improved visual comfort while protecting the lamella from outdoor weather conditions. Other opportunities are available using conventional c-SI cells or organic photovoltaic modules.

To effectively utilise dynamic shading systems, intelligent algorithms are essential to orientate blinds for optimal energy efficiency. These algorithms help find the balance between PV generation and daylight control, minimising heating, cooling, and lighting demands.

Another solution is user-operated low-tech systems, which follow traditional architectural patterns. The "Solar Window Shutter" (Figure 3.15) is a product developed and patented by the German architect Astrid Schneider [11].

The "Solar Window Shutters" can be moved by hand to either fully shade, be in a ventilation position, or on the side to allow full sunshine into the room, while producing solar electricity. Due to a parallel opening mechanism, the active solar cell side is always oriented towards the sun so that the inhabitant can choose freely and depending on the actual weather conditions and their needs between fully open or fully closed shutters.

Figure 3.13 Semi-transparent thermally insulated BIPV façade in an office building. Photo: Sunovation GmbH, Germany.

Figure 3.14 The movable BIPV solar shading system, Solskin, installed at HiLo, NEST building, in Dübendorf. Photos: courtesy of Roman Keller (left) and Chair of Architecture and Building Systems (A/S) Group (right).

Figure 3.15 Residential building in Nechlin, Germany, with "Solar Window Shutters" designed and patented by Astrid Schneider. Photo: Astrid Schneider, Solar Architecture, Berlin.

In conclusion, semi-transparent and translucent solar façade concepts with solar cells integrated into the façade glazing itself have limitations in adapting to dynamic solar conditions. However, the integration of PV modules into movable shading systems offers opportunities to combine shading benefits with solar energy harvesting. Lightweight PV materials like CIGS provide flexibility and enable fine-tuning of functions while ensuring visual and daylight comfort. Innovative concepts, including external movable shading and integrated windows, control sunlight and solar gains, contributing to both energy efficiency and architectural appearance. Intelligent algorithms play a crucial role in optimising the performance of dynamic shading systems by balancing PV generation and daylight control.

3.3.4 Other BIPV applications and designs

From innovative façades or roofs that generate electricity to elegantly designed shading devices discussed in the previous chapters, BIPV is redefining the boundaries of architectural creativity and sustainability. Some other BIPV applications and designs have been developed to push forward renewable energy production in buildings. These possibilities include:

Balustrades: BIPV modules can be integrated into balustrades or guardrails, providing safety and generating solar energy simultaneously. These solutions often prefer bifacial technologies to harvest sunlight from both sides (Figure 3.16).

Parapets: Solar modules can be incorporated into parapet walls, which are low protective walls at the edge of a roof or balcony. This application combines aesthetics with energy generation, using surfaces that normally protrude from the façade of the building.

Winter gardens: BIPV can be incorporated into winter garden structures, providing a dual function of growing plants and generating energy.

Pergolas and gazebos: BIPV modules can be integrated into pergolas and gazebos, offering shade and energy generation in gardens and recreational areas.

Figure 3.16 From balustrades to solar benches, BIPV can take several forms within the built environment. Photos: courtesy of Isa Zanetti (left) and Costa Kapsis (right).

Solar benches and tables (Figure 3.16): These elements can integrate BIPV modules and thus provide seating and device charging capabilities.

Shading elements for public spaces: Due to the ongoing climate change, more and more cities want to provide shading elements for public open spaces.

Finally, different **artistic designs** can be proposed where BIPV can be used in artistic or decorative installations, allowing architects and designers to create unique and functional art pieces.

References

[1] C. Ballif, F.-J. Haug, M. Boccard, P. J. Verlinden, and G. Hahn, "Status and perspectives of crystalline silicon photovoltaics in research and industry," *Nat. Rev. Mater.*, vol. 7, no. 8, pp. 597–616, Mar. 2022, doi:10.1038/s41578-022-00423-2.

[2] A. Metz, M. Fischer, and J. Trube, "International technology roadmap for photovoltaics (ITRPV) 8th edition: Crystalline silicon technology-current status and outlook," in *Workshop ITRPV – VDMA Photovoltaic Equipment*, ITRPV, 2017, pp. 1–40, https://etip-pv.eu/publications/etip-pv-publications/download/international-technology-roadmap-for-photovoltaics.

[3] SUPSI, "Standardization, performance risks and identification of related gaps for a performance-based qualification in BIPV," 2019 [Online]. Available: https://bipvboost.eu/public-reports/download/standardization-performance-risks-and-identificati.

[4] A. M. Law, L. O. Jones, and J. M. Walls, "The performance and durability of Anti-reflection coatings for solar module cover glass – a review," *Sol. Energy*, vol. 261, pp. 85–95, Sep. 2023, doi:10.1016/j.solener.2023.06.009.

[5] G. Eder, et al., "Coloured BIPV market, research and development IEA PVPS task 15, Report IEA-PVPS T15-07: 2019," 2019 [Online]. Available: http://iea-pvps.org/index.php?id=task15.

[6] P. Bonomo, et al., "Categorization of BIPV applications. IEA-PVPS T15-12:2021," 2021 [Online]. Available: https://iea-pvps.org/wp-content/uploads/2021/09/IEA-PVPS-T15-12_2021_BIPV-categorization_report.pdf.

[7] F. Frontini, P. Bonomo, D. Moser, and L. Maturi, "Building integrated photovoltaic facades: Challenges, opportunities and innovations," in *Rethinking Building Skins*. Elsevier, 2022, pp. 201–229. doi:10.1016/B978-0-12-822477-9.00012-7.

[8] W. Bahr, "A comprehensive assessment methodology of the building integrated photovoltaic blind system," *Energy Build.*, vol. 82, pp. 703–708, 2014, doi:10.1016/j.enbuild.2014.07.065.

[9] M. Mandalaki, K. Zervas, T. Tsoutsos, and A. Vazakas, "Assessment of fixed shading devices with integrated PV for efficient energy use," *Sol. Energy*, vol. 86, no. 9, pp. 2561–2575, Sep. 2012, doi:10.1016/j.solener.2012.05.026.

[10] X. Zhang, S.-K. Lau, S. S. Y. Lau, and Y. Zhao, "Photovoltaic integrated shading devices (PVSDs): A review," *Sol. Energy*, vol. 170, pp. 947–968, Aug. 2018, doi:10.1016/j.solener.2018.05.067.

[11] T. Reijenga, et al., "Successful building integration of photovoltaics – A collection of international projects," *International Energy Agency*, 2020 [Online]. Available: https://iea-pvps.org/wp-content/uploads/2021/03/IEA-PVPS-Task-15-An-international-collection-of-BIPV-projects-compr.pdf.

[12] https://www.bccresearch.com/pressroom/egy/global-bifacial-solar-market-set-to-surge-reaching-1871-billion-by-2028

[13] IEA (2023), World Energy Outlook 2023, IEA, Paris https://www.iea.org/reports/world-energy-outlook-2023, Licence: CC BY 4.0 (report); CC BY NC SA 4.0 (Annex A).

CHAPTER 4

A decision-making process for BIPV design

Costa Kapsis, Rebecca Yang, Pabasara Wijeratne, and Nilmini Weerasinghe

4.1 Introduction

BIPV design decisions encompass a multi-objective optimisation process. Depending on the BIPV application, certain objectives may conflict with each other (for instance, aesthetics versus efficiency). These dynamics create a network of interconnected variables, which can complicate the decision-making process for designers.

This chapter consolidates existing knowledge about PV and BIPV design and, whenever possible, strives to simplify the decision-making process into a sequence of steps that assists designers in streamlining the design workflow while mitigating risks and uncertainties [1,2]. The workflow should follow an integrated design process between the architect, the BIPV consultant, the electrical, mechanical, and structural engineers, the BIPV installer, and the client. These steps aim to guide designers towards nearly optimal design solutions based on well-defined criteria, spanning from the preliminary design phase to construction. Initially, this chapter explores the influence of essential design parameters such as climate, location, orientation, and shading through a step-by-step design workflow. The final section outlines the available business models, from net-metering to BIPV-as-a-service. Operation, safety, and maintenance of BIPV systems is discussed under Chapter 6.

A robust path for designing high-performance, sustainable, and occupant-centric buildings mandates three steps: energy conservation, energy efficiency, and renewable energy generation [3].

Energy conservation refers to the reduction of building energy consumption through passive means. Key principles of energy-conserving design include an airtight, well-insulated building envelope, moderate window-to-wall ratio (i.e., 50% or less), high-performance windows to reduce heat exchange with the outdoors while providing sufficient daylight, solar shading to reduce solar heat gains during cooling season, exposed thermal mass to regulate operative temperatures, and the adoption of natural ventilation techniques.

Energy efficiency refers to the reduction of building energy consumption through the adoption of energy-efficient equipment such as high-performance heat pumps and heat (or enthalpy) recovery ventilation, LED lighting, and active storage (thermal or electrical). Advanced building controls that can reduce and shift peak load demand are also considered *energy efficiency* measures.

Finally, **renewable energy generation** through a BIPV (or PV) system should be considered after energy-conserving and efficient building design measures have been implemented to reduce the energy loads and peak demand as much as possible and in a cost-effective manner. Given that BIPV is a building envelope technology, there is a need for harmonising energy-conserving architecture with near-optimal solar electricity generation.

The following steps provide an easy-to-implement decision process and workflow for the effective implementation of BIPV. Figure 4.1 illustrates the BIPV design workflow relative to the design and construction stages of a building.

DOI: 10.1201/9781003432241-4

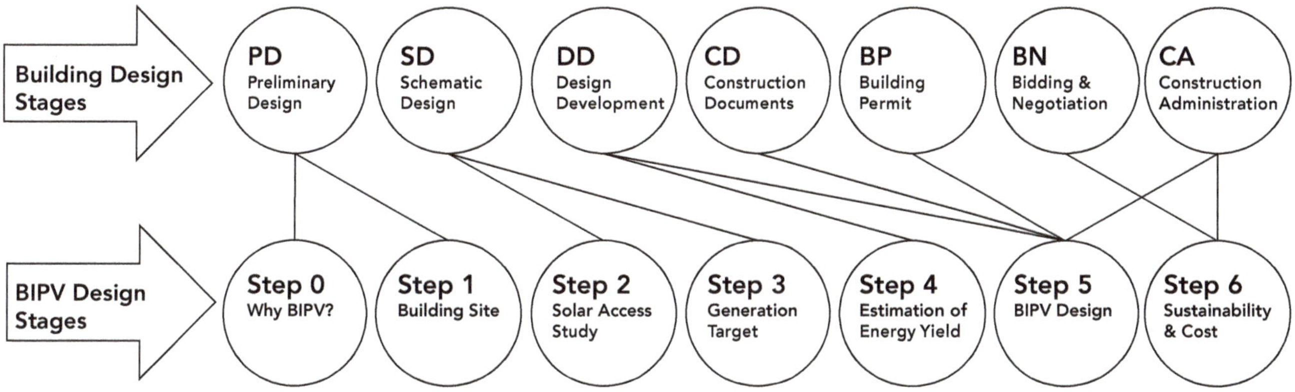

Figure 4.1 BIPV design workflow throughout the building design stages.

4.2 Step 0: Why BIPV?

BIPV can create a compelling value proposition by transforming traditional building surfaces into on-site power-generating assets. Several key benefits can resonate with various stakeholders and make BIPV an attractive investment [4]:

Energy Generation and Savings: While the initial investment in BIPV may be higher compared to traditional cladding materials, its long-term value stems from energy savings through reduced electricity bills and potential revenue generation through excess energy production.

Regulatory Compliance: BIPV installations can align with local regulations and sustainability standards, positioning the building for compliance with evolving energy efficiency requirements (e.g., net-zero energy), and enabling the transition towards the electrification and decarbonisation of buildings.

Enhanced Aesthetics: From heritage to contemporary architecture, BIPV can blend seamlessly with the building design, while "camouflaging" a PV module within the form of a window, roof tile, or coloured cladding that satisfies both technical and aesthetic requirements.

Increased Property Value: Buildings equipped with BIPV systems create the potential for higher property values due to reduced operating costs, sustainability, and resilience features.

Corporate Green Status (if applicable): The use of BIPV demonstrates a commitment to sustainability and innovation by reducing carbon emissions and reliance on non-renewable energy sources, which can enhance an organisation's brand image and reputation among customers, partners, and stakeholders.

Renewable Energy Credits and Incentives (if applicable): Many regions offer financial incentives, tax credits, and rebates for PV installations, including BIPV systems. These incentives can significantly lower the upfront costs and improve the return-on-investment.

4.3 Step 1: Assess building site

At the preliminary design stage, an on-site visit and site survey are essential for better understanding the solar access potential. A site that has a relatively unobstructed sky view toward the south ± 45 degrees (north ± 45 degrees, for the southern hemisphere) can be a good candidate for a BIPV façade and roof application. When surrounding buildings and vegetation substantially obstruct the sky view, a BIPV roof might be the preferred option.

4.3.1 Orientation and tilt angle

The following recommendations should be treated as rules of thumb. Further assessment should be performed using building performance simulations at different stages of the project design, to account for the local climate, building use, form and orientation, routine shading (e.g., self-shading, surrounding buildings, vegetation, topography), and temporary shading (e.g., deciduous vegetation). It is important to consider the changing energy prices and potential high-yield periods, which can make BIPV installations valuable and cost effective even in areas with lower annual energy yield.

For unobstructed near equator-facing BIPV systems, tilt angles equal to latitude ±15 degrees will ensure high annual energy yield. In Madrid, Spain, where the geographic latitude is ~40°, an unobstructed BIPV roof, skylight, or solar shading device at a tilt angle between 15° and 55° will ensure high annual electricity generation. To optimise seasonal energy generation, a BIPV tilt angle near latitude −15 degrees will yield peak electricity generation over the winter months, while a tilt angle near latitude +15 degrees will peak during the summer season (Figure 4.2).

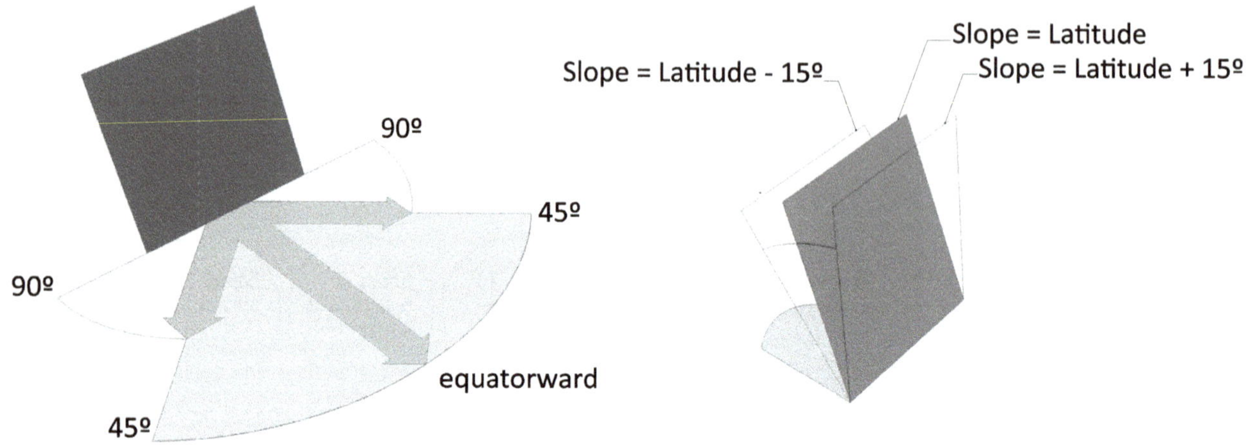

Figure 4.2 Rules of thumb for BIPV system orientation (left) and tilt angle (right).

BIPV roof applications are well-suited for lower geographic latitudes but can also be cost-effective and beneficial in mid to high latitudes. These systems can be oriented from southeast to southwest (northeast to northwest for the southern hemisphere), with the highest energy yield achieved for near equator-facing roofs.

BIPV roof applications are ideal for locations with mid to high geographic latitudes but can also be installed in other regions. They can be oriented from east to west, with the highest energy yield occurring for near equator-facing façades.

For solar shading applications, horizontal BIPV overhangs or louvers are recommended for near equator-facing windows while vertical louvers will be most effective (in terms of electricity generation and solar shading control) for near east- and west-facing windows. The size, tilt angle, and louver spacing can be determined through a solar shading analysis. Note that inter-louver shading should be avoided during periods of high solar flux.

4.3.2 Utility requirements

Local utilities may have specific requirements or guidelines for (BI) PV installations. It is essential for the design team to be aware of any size, metering, and grid connection constraints early on. The BIPV design should be optimised to achieve project objectives while complying with any limitations imposed by the local utility.

4.3.3 Zoning bylaws

It is important to get familiar with the local zoning bylaws and building codes as they can impact the design and output of the BIPV system. Requesting access to building permit records for neighbouring construction sites can help the designers foresee future conflicts and design to mitigate limited solar access, e.g., an upcoming neighbouring construction can cast shadows on the BIPV system, resulting in a substantial reduction of annual electricity generation. Note that local bylaws may protect the right to sunlight access through urban planning guidelines or solar easements. Through a solar easement, the property owner gains legal protection and assurance that their solar access will be maintained over time.

4.3.4 Is it a heritage designation?

Buildings located within heritage conservation districts (or designated as heritage properties) may be subject to restrictions that limit the installation of PV systems. Many of these bylaws were initially developed to avoid the use of conventional rooftop PV systems and the aesthetics associated with them. If BIPV module products tailored for heritage applications are incorporated (e.g., terracotta solar tiles), the limiting bylaws may not apply or can be disputed, as the BIPV system will be seemingly blended within the historic character of the building or the district.

4.4 Step 2: Perform a solar access study

Considering BIPV at the schematic design stage of new construction or renovation can impact the overall success, functionality, and sustainability (environmental, social, and economic) of the project.

4.4.1 Solar availability

As described in Chapter 2, the quantity and quality of sunlight incident on the surface of the BIPV installation serves as the primary determinant of its energy output: higher solar flux (e.g., clear sunny days) leads to increased electricity generation, while reduced flux (e.g., overcast days) results in reduced output. PV modules generate electricity primarily from direct sunlight (also known as direct or beam solar irradiance), but both direct and diffuse sunlight play an important role in their overall performance. To ensure high energy yield, a BIPV system should be designed to maximise annual exposure to direct sunlight.

Geographic latitude influences solar availability, with seasonal variations increasing at higher latitudes due to the Earth's orbit around the sun. Generally, summer seasons tend to have higher solar flux due to longer days compared to autumn, spring, and winter seasons. However, local climate, weather, topography, and terrain can significantly affect solar availability and seasonal variations. Locations with high humidity, air pollution, frequent cloud cover, and precipitation will experience reduced solar radiation. The presence of mountains, hills, buildings, and trees can cast shadows, reducing the amount of sunlight that reaches the surface of the solar modules. On the contrary, flat, open landscapes maximise solar exposure. It is essential that the designers become familiar with the local climate and evaluate and quantify solar availability through a solar access simulation study.

4.4.2 Solar access simulation study

For new construction, an annual solar access study, in conjunction with the building massing study, can guide the architect to optimise the building form for solar accessibility and reveal major limitations and opportunities. For example, for a building located near the city centre, the proximity of other tall buildings, located south (north for the southern hemisphere) of the east-west axis of the building, might lead to significant shading and a reduction of available surfaces for BIPV. The intention of a solar access simulation study is not to size the BIPV system but rather to visualise and quantify the distribution and intensity of solar energy on building surfaces while accounting for shading from surrounding buildings, vegetation, and the building itself (i.e., self-shading).

For a building retrofit, a solar access simulation study can help identify the building surfaces suited for BIPV.

Depending on the simulation tool, the solar access study will provide a heatmap for each surface of the massing model reporting the solar exposure (in kWh/m^2 or sun-hours[1]) per month or annually (Figure 4.3). Typically, a solar access study accounts only for direct sunlight, excluding diffuse irradiation from the sky or surrounding environment.

To perform a solar access simulation study for the project site, a year hourly weather file is required as an input. Various national and international platforms exist that provide weather files, for free or for a fee, developed based on historical weather data, satellite data, or statistical models [5]. The following international free databases can be used:

- PVGIS: https://re.jrc.ec.europa.eu/ (for typical meteorological weather files)
- Climate.OneBuilding: https://climate.onebuilding.org/ (for typical and future meteorological weather files)

Note that selecting a suitable weather data source can be challenging for remote locations or when no nearby weather station is available for the project site. To address this, the designers can start by identifying candidate sources with similar latitude and elevation. Comparing monthly statistics between candidate sources and the project site can help designers identify an adequate match.

4.5 Step 3: Determine annual BIPV energy generation target

While a conventional PV system is typically designed to maximise annual energy yield, a BIPV design prioritises building performance and building-grid interaction. Initially, the annual building energy consumption can be estimated using building performance simulation (BPS) software that is able to estimate the building energy demand for heating, cooling, lighting, plug loads (including electric vehicles), and account for stationary electric storage, occupancy behaviour, and building controls. Beyond the estimation of the annual building energy consumption, it is important that the building consumption profile and peak power demand are also taken into consideration to minimise energy-related costs and emissions. Commonly, a BIPV system is designed to meet more than one of the following building energy requirements.

4.5.1 Partial energy offset target

At its most basic, an energy generation target can be determined as a fraction of the annual building energy consumption; e.g., if a new house is estimated to consume 10,000 kWh per year and a BIPV roof is estimated to generate 1,100 kWh/kW, then the size of the BIPV roof to cover 60% of the annual building consumption is estimated as follows:

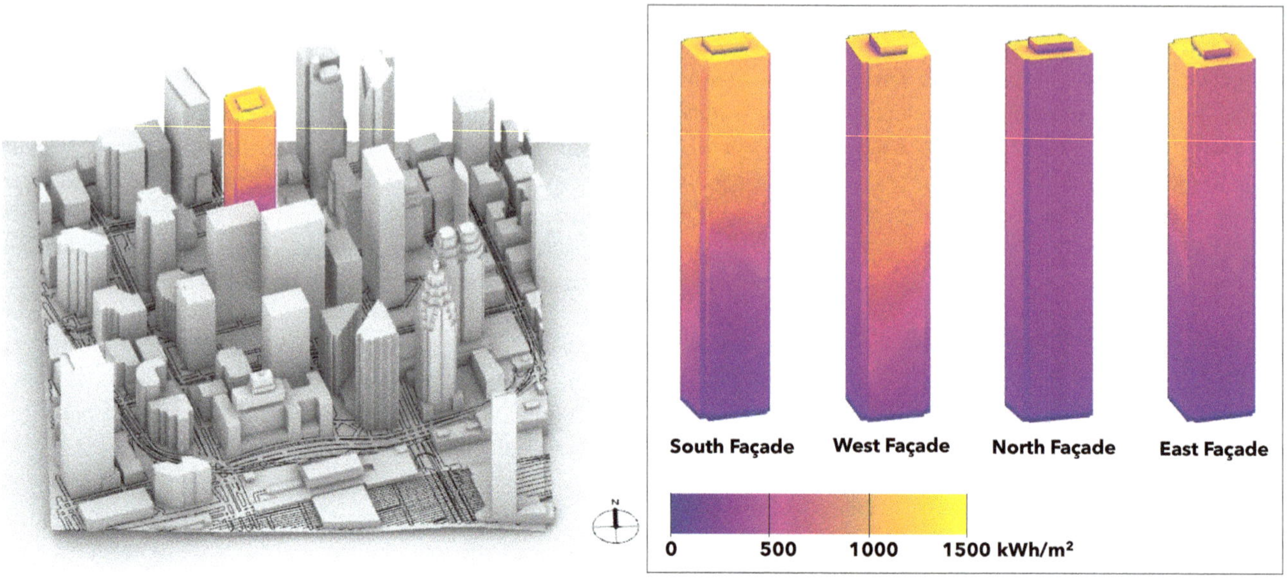

Figure 4.3 Example of a solar access study accounting for the surrounding buildings, downtown Toronto, Canada.

60% × 10,000 kWh ÷ 1,100 kWh/kW ≈ 5.5 kW of BIPV roof. While this simplistic approach does yield some energy cost savings, it does not account for any building-grid interactions or dynamic electricity pricing and, thus, it is far from optimal.

4.5.2 Self-consumption target

To minimise building energy costs and maximise self-consumption, the BIPV energy generation profile should closely match the building consumption profile. For commercial and institutional buildings where the peak demand occurs around noon, a near south-facing BIPV system is recommended. On the other hand, near east- and near west-facing BIPV systems can assist with offsetting morning and afternoon peak power demand for residential buildings. The BIPV orientation and installed capacity should be selected to provide the highest self-sufficiency index (*SSI*), while the self-consumption index (*SCI*) is at 100% [6].

4.5.3 Net-zero energy target

Although several definitions exist, a net-zero energy building (NZEB or NetZEB) can be defined as a building that produces as much on-site energy as it consumes in an average year. In the case of a fully electric building, the load/generation energy balance at the utility meter is [7]:

$$\sum_i E_{PV_i} - \sum_i \left(E_{H_i} + E_{C_i} + E_{L_i} + E_{HW_i} + E_{PL_i} + E_{EV_i}\right) = E_{PV} - E_{LOADS} \geq 0 \quad (4.1)$$

where E_{PV} and E_{LOADS} stands for annual photovoltaic generation and electric energy consumption, respectively, and the summation (Σ) covers a complete year. The BIPV system should be sized to meet or exceed the annual electric energy consumption (kWh/year or kWh/m²/year) for heating E_{H_i}, cooling E_{C_i}, lighting E_{L_i}, hot water E_{HW_i}, plug loads E_{PL_i} and potentially electric vehicle charging E_{EV_i}. Building-grid interaction and primary energy (i.e., source energy) are not considered in this energy balance. For optimal performance, the building should be designed and operated to maximise SCI, using demand-side management (DMS) strategies.

4.5.4 Carbon-neutral target

A carbon-neutral building can be defined as one whose construction and operation do not contribute to emissions of greenhouse gases (GHG) over its life cycle. In this case, the BIPV system should be sized to offset embodied and operational carbon emissions, effectively resulting in a net-positive energy building. The emission balance can be calculated as follows:

$$\sum_i \left(GHG_{embodied+,i} + GHG_{operation+,i}\right) - \sum_i GHG_{BIPV-,i} \leq 0 \quad (4.2)$$

where the summation (Σ) refers to building life cycle (i.e., 50 years or more). Note that GHG_{BIPV-} refers to GHG emissions offset by BIPV while $GHG_{embodied+}$ and $GHG_{operation+}$ refer to GHG emissions from production (A1-A3), construction (A4-A5), use (B1-B5), and end of life (C1-C4) of the building. For a fully electric building, if equation 4.1 is equal to zero, then the building meets operational carbon neutrality.

4.6 Step 4: Estimate BIPV installed capacity and energy yield

At the design development stage, the BIPV consultant – in close coordination with the architect to maintain or enhance aesthetics – should determine what might be the most effective BIPV configurations (e.g., roof, façade, solar shading) to satisfy the annual BIPV energy generation target from Step 3. Knowing the available surface area (e.g., for BIPV roof), viable solar capacities and electrical output of the BIPV system can be estimated using PV simulation tools. Most of these tools are designed to estimate the yield of PV ground-mounted systems and can conduct shading analysis with varying degrees of accuracy. However, PV software are not able to capture any interactions between the BIPV system and the building (Figure 4.4). In the case of a BIPV roof or rainscreen, the use of PV simulation software could be sufficient, if the BIPV system has little to no impact on building heating, cooling, and lighting loads [8].

On the other hand, BIPV curtain wall, skylight, and solar shading applications can impact daylighting, solar heat gains, and visual and thermal comfort. This, in turn, affects the building's heating, cooling, and electric lighting loads. Therefore, using simulation software is advised. Building performance simulation tools are able to evaluate the BIPV's effect on building energy performance and occupancy comfort. A parametric analysis (e.g., varying the packing density of solar cells or transparency of the thin-film layer) can assist the building designer to capture these effects and optimise the design of the BIPV system. Simulation tools can also often estimate BIPV electrical output, removing the need for separate PV software. Alternatively, the BIPV system hourly power output (P, in watts) can be approximated using Evan's simplified model [9]:

$$P = PR \cdot \eta_{STC} \cdot \left(1 - L_{temp}\right) \cdot G \cdot A_m \tag{4.3}$$

where η_{STC} is available in the BIPV manufacturer's product specification sheet or derived from equation 2.1, L_{temp} is determined using equation 2.2, *PR* is estimated using equation 2.3, and *G* is computed through simulation tools. If the BIPV system has multiple arrays with varying orientations or tilt angles, it is recommended to compute the electrical output for each array independently. The total hourly power output of the entire BIPV installation (*P*) is the sum of each hourly array output power.

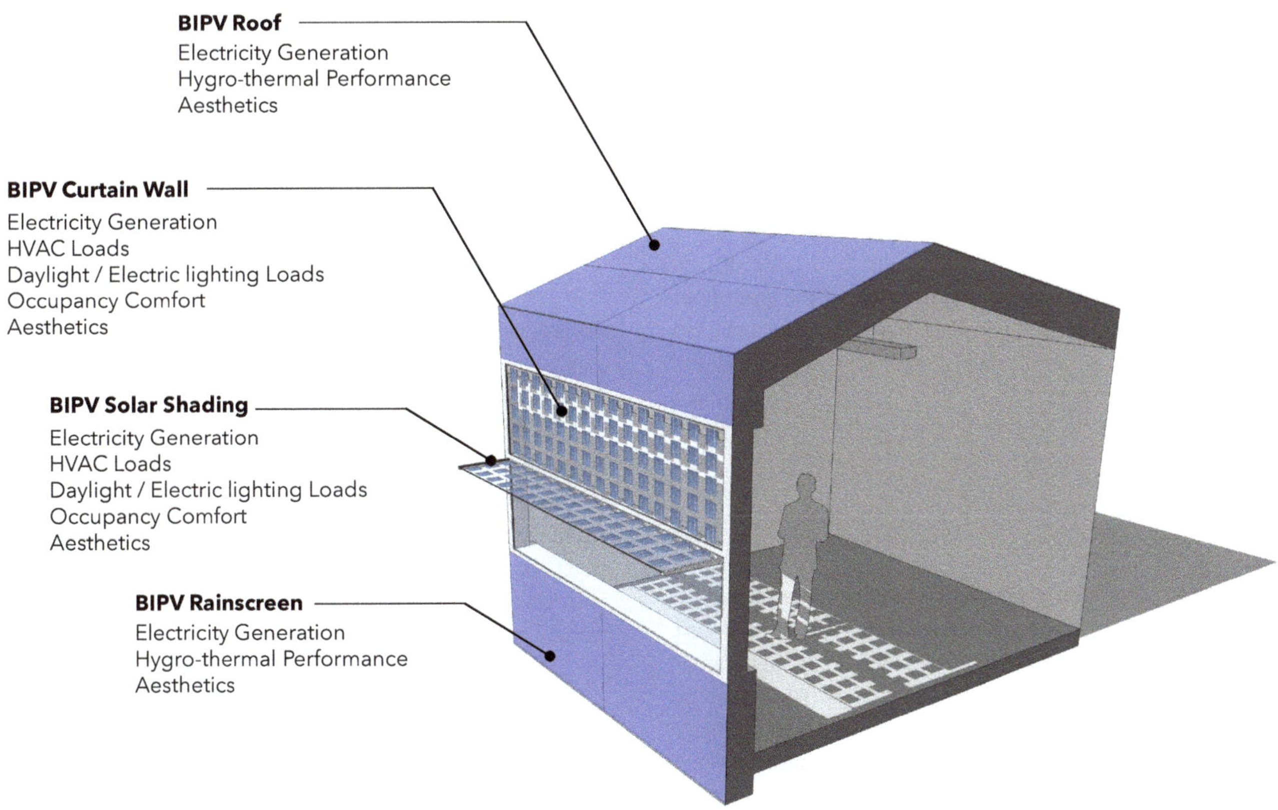

Figure 4.4 Major interactions between BIPV application, building systems, and occupants should be assessed and optimised using building performance simulations. Adapted by NRCan.

The estimated energy output of the system over a period (E_{PV}, in kWh) is given by:

$$E_{PV} = \sum_{i} P_i \cdot \Delta t \tag{4.4}$$

where usually Δt = 1 hour, and the summation (Σ) covers a complete year. Most PV software tools use complex electrical models such as the equivalent one-diode [10] and the Sandia [11] model to estimate PV power values. Typically, each software offers a database of PV modules and inverters (or provides user-defined options), allowing users to make selections based on the system design. Advanced inverter simulation models are available, designed to accurately capture AC system performance under various design and operational conditions.

4.6.1 Shading

Ideally, a BIPV array should be shadow-free at periods of high solar flux, to maximise yield and avoid potential module damage due to hotspots. Yet, in urban or suburban settings, some shading is inevitable. Whether the shading is routine (e.g., self-shading or from surrounding structures) or temporary (e.g., vegetation, snow, or dirt), it can be estimated using PV or building simulation software. Most software tools can capture routine shading. Various shading models exist, with most of them being some form of a raytracing technique (e.g., polygon clipping, pixel counting, backward raytracing). The shading on the BIPV surface is either calculated in every simulation time-step (e.g., hourly) or using a greater length of time (e.g., monthly), with the latter being less time consuming but also less accurate. Shading mitigation techniques are discussed under Step 5. For temporary shading such as vegetation, most software rely on empirical models or user-defined monthly values. Evergreen trees casting shadows on a BIPV installation are typically simulated as opaque objects. However, simulating shading from deciduous trees is more complex due to their changing opacity throughout the seasons.

4.6.2 Soiling, snow, and ice

Capturing shading effects caused by soiling is challenging due to their dependence on the installation, site, and local environmental conditions. Common causes include mineral dust (e.g., in arid locations), air pollution (e.g., in metropolitan or industrial centres), pollen, and bird droppings. When shading from soiling is uniform, it primarily reduces the system's electricity yield. However, localised soiling, such as bird droppings, fallen leaves, or graffiti, can create module hotspots, posing a risk of potential damage to the module.

Soiling can reduce monthly power output from 2% to 17% in extreme cases [12]. However, in most systems, soiling-induced losses can be assumed to be less than 5%. Occasionally, a "dirt dam" can form at the bottom edge of framed BIPV modules, potentially causing localised shading. This is not an issue for frameless products. BIPV systems with tilt angles of 15° or more tend to clean during rainfall. In regions prone to heavy soiling, it is recommended to clean BIPV modules at least once a year. The use of hard water should be avoided as it can leave water spots and result in calcium build-up on the BIPV module surface, leading to scaling.

In regions prone to snow, PV systems often experience electricity generation losses due to snow accumulation. While annual electricity generation losses can be assumed to be less than 10% in most climates, monthly losses can exceed 25% during peak winter months, mainly due to formation of ice that prolongs snow retention on the module surface [13]. The use of frameless BIPV products at a slope of 30° or steeper can significantly minimise snow accumulation.

4.7 Step 5: Develop the BIPV design

The BIPV system design should coincide with the heating, ventilation, and air conditioning (HVAC) system and building envelope design, spanning across the stages of design development and construction documents. This section offers overarching guidelines for grid-connected BIPV system design. For detailed design requirements specific to various BIPV applications, such as roofing, façades, and external integration (e.g., solar shading), readers are directed to Chapter 5.

4.7.1 Modules

The choice of module size, colour, efficiency, and other features, such as solar cell technology and the decision between framed versus frameless, should be determined based on system configuration (e.g., rainscreen vs. curtain wall), desired aesthetics, and required system capacity. When possible, the number of module variations should be minimised to reduce manufacturing and installation costs. In areas where shading might be an issue, dummy modules are recommended to maintain a uniform appearance. Additionally, the modules must be compatible with the selected mounting system.

4.7.2 Mounting system

The mounting system should be engineered and installed to withstand relevant static, environmental, and live structural loads. In addition, the modules should be mounted to accommodate: (1) thermal expansion and (2) adequate air flow. The mounting system and connections should be engineered with future reuse in mind. When applicable, the connections between the building structure and the BIPV mounting substructure should be installed to accommodate differential movement.

4.7.3 Strings and arrays

Modules that are exposed to similar irradiation should be connected in series to form a PV string. Multiple PV strings connected in series and parallel configurations form a PV array. The PV string length is determined by the inverter input voltage limits. For instance, a module with an open circuit voltage of V_{OC} = 38 V at STC and a voltage temperature coefficient of $\beta_{VOC} = -0.3\%/°C$ will have a voltage of $38\ V \cdot [1 + (-0.3\%/°C) \cdot (0°C - 25°C)] = 40.9\ V$ at 0°C. A string length of 20 modules will have an open circuit voltage of 817 at 0°C and can be safely connected to a string inverter with a maximum input DC voltage of 820 V. In this example, 0°C refers to the lowest design outdoor temperature that is dependent on the building location.

4.7.4 Balance of system

Balance of system refers to all the components of a PV system other than PV modules, which include power conversion equipment (inverters, DC-DC optimizers), protection devices, cabling, etc. Grounding, AC and DC cabling size and management, and the implementation of a rapid shutdown switch are specified by the local electrical codes. Using cable trays, conduits, clips, and ties can secure wiring, reduce mechanical stress, prevent abrasion, and guard against damage from wind or vibrations. All strings should be wired to prevent induction loops, which can generate magnetic fields [14]. When possible, the string modules should be connected in a leapfrog fashion vs. daisy chain wiring to reduce DC wiring costs (Figure 4.5).

4.7.4.1 Inverters

Grid-connected inverters can be categorised based on grid-interaction capabilities and PV connection configuration (Figure 4.6). Traditional grid-connected inverters align with the grid's waveform frequency and automatically shut off during a grid power outage. On the other hand, grid-forming inverters will also disconnect from the grid during a power outage, but they will continue supplying the building with electricity (i.e., island mode). Typically, grid-forming inverters are able to provide charge controller capabilities, allowing the charging/discharging of a stationary electrical storage (e.g., battery bank). At minimum, the electrical storage should be sized to supply emergency loads. Upon utility request, grid-forming inverters can also offer grid support services, such as enhancing voltage or managing reactive power. Their use is recommended when building energy resilience is a priority as they can continue feeding the building or the microgrid with solar electricity.

When it comes to PV connection configurations, the BIPV strings can be connected to central, string inverters (multi-string inverters), or microinverters (Figure 4.7). The additional use of DC-DC optimizers can help reduce energy losses and hotspots due to shading variations between string modules. Whether central or string, inverters should be installed in locations, either indoors or outdoors, that are safe from flooding and always meeting the operating condition range required by the manufacturer. A decentralised connection option involves using microinverters, which optimises the performance of intricate BIPV systems. Each microinverter typically connects to one or

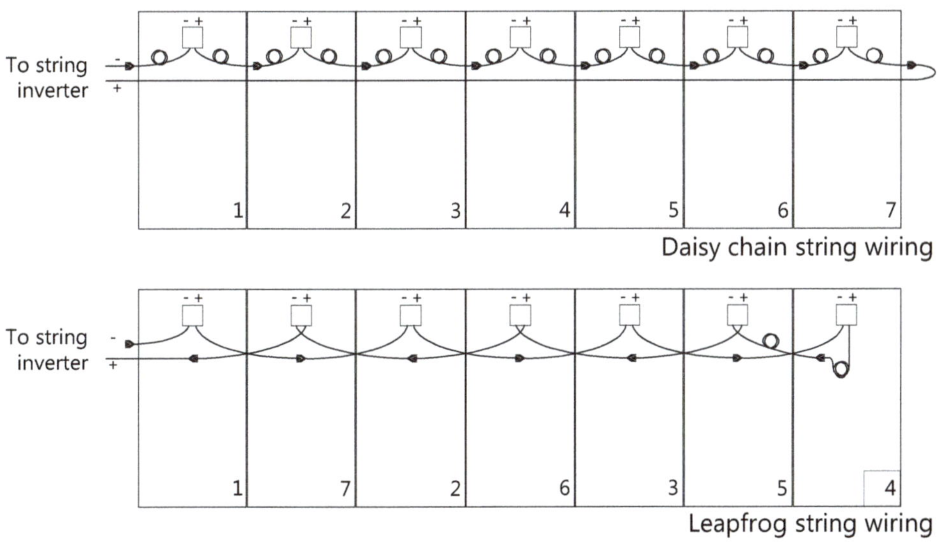

Figure 4.5 Leapfrog vs. daisy chain wiring to reduce DC wiring for a string.

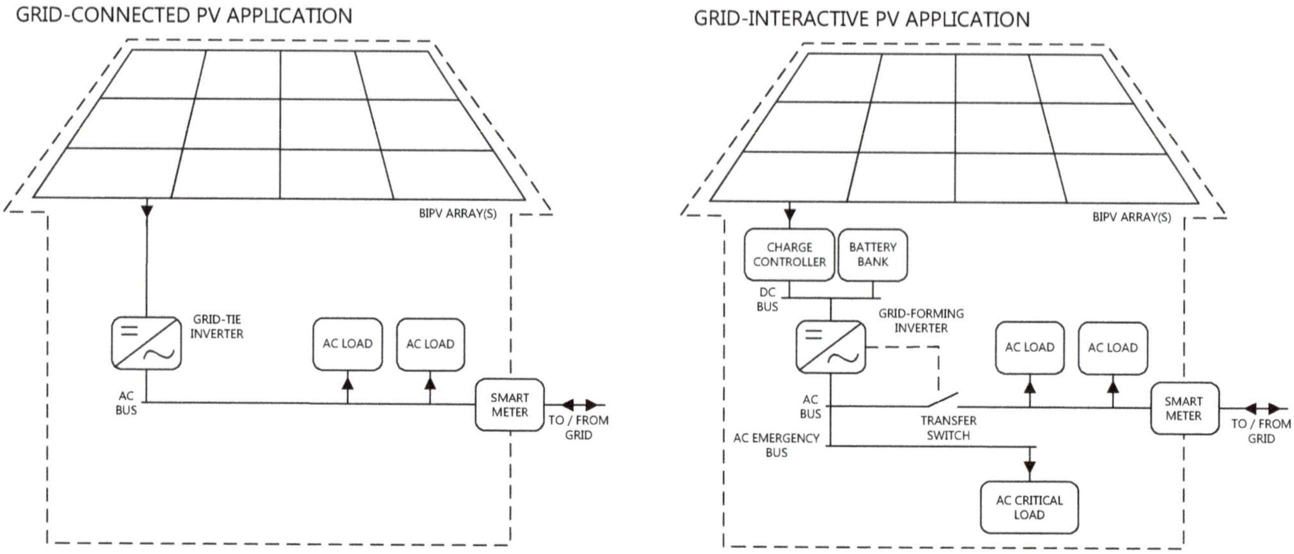

Figure 4.6 Two common grid-connected BIPV system configurations.

Copyrights: NRCan.

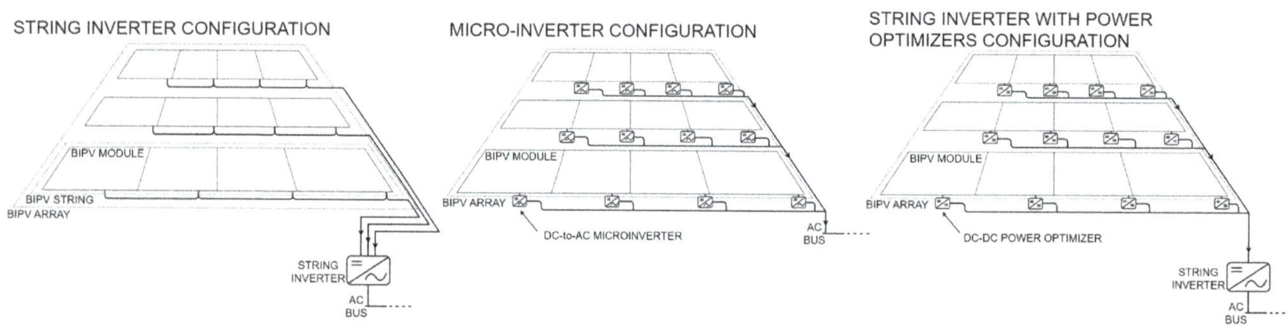

Figure 4.7 Three common PV connection configurations for BIPV systems.

Copyrights: NRCan.

two modules and is often positioned behind the module, which can make maintenance more challenging.

4.7.5 Apply for utility interconnection

At this stage, it is important to apply for an interconnection permit because local utility requirements and regulations might limit BIPV installed capacity. In other cases, local electric utility feeders may not support additional generation. Once the local utility grants approval for the installation, the installation process can be initiated.

After the BIPV installation is completed, a local utility inspector will assess the system to ensure it complies with electrical and fire safety codes. Subsequently, a smart meter is installed to facilitate two-way electricity flow between the building and the grid. Upon a successful inspection, the inspector issues a permission to operate (PTO) authorisation.

Depending on local codes and regulations, a construction permit might also be required for both new and retrofit BIPV projects. Exemptions from building permits may be available for BIPV installations below a certain capacity threshold.

4.8 Step 6: Assess sustainability, circularity, and life cycle cost

A BIPV system consists of several materials, products, and components that have operational life cycles spanning from

15 (e.g., string inverters) to 100 years (e.g., BIPV array supporting structure). The current linear design approach to building integration tends to consider demolition and recycling as the only end-of-life option for BIPV systems. However, an upfront investment in a sustainable and circular system design can lead to significant long-term savings in terms of energy costs, maintenance, and replacements. The envelope's durability and performance over time can result in reduced operational expenses. This section attempts to provide initial guidelines and metrics to evaluate the sustainability and circularity of a BIPV system at the early design stage.

4.8.1 Non-renewable energy payback time

A simple but effective key performance indicator (KPI) to assess the environmental impact of a BIPV system is the non-renewable energy payback time, which refers to the time it takes for the BIPV system to "pay off" the primary energy used to manufacture the system itself. Table 4.1 presents this simple metric for an average residential PV system, accounting for the life cycle analysis (LCA) of PV modules, wiring, mounting structure, inverter, and system installation, with a service life of 30 years (15 years for the inverters). These values are also appropriate for BIPV roof and rainscreen systems but may vary for other applications such as BIPV curtain walls and canopies or for BIPV installations that use glass-glass and/or coloured modules. It is worth noting that BIPV is the only building envelope technology that offers an energy payback time as it has the unique characteristic of renewable energy generation. As PV efficiencies and manufacturing continue to improve, the non-renewable energy payback time for BIPV is expected to reduce.

4.8.2 Environmental footprint

Non-renewable energy payback time is just one aspect of evaluating the LCA of a BIPV system; other factors such as carbon emissions, resource use, and water usage also play an important role in making informed design decisions. Table 4.2 provides a list of environmental KPIs for a PV system, aiming to support a preliminary LCA at the schematic design or a comparative LCA at the design development stage of the building to (1) inform design decisions, (2) support sustainability certification and upcoming regulatory changes (e.g., carbon tax), and (3) enhance business-to-business (B2B) and business-to-client (B2C) communications. These values should be used when no values are made available for specific markets/countries or through manufacturers/installers. A more accurate LCA can be performed after the construction procurement stage, using the actual BIPV configuration and product quantities.

Table 4.1 Non-renewable energy payback time KPI, for an average grid-connected residential PV system, using framed glass/glass PV modules [15].

	Crystalline Silicon PV system		Thin-film PV system	
	m-Si	mc-Si	CdTe	CI(G)S
Non-renewable energy payback time (years)	1.2	1.2	0.9	1.3

4.8.3 Circularity

A BIPV system should be designed with reversibility in mind. An upfront investment in a circular BIPV system design can lead to significant long-term savings in terms of maintenance, replacements, and upgrades. The system's durability and performance over time can result in reduced operational expenses.

Fastening Methods: Design for disassembly using reversible fastening methods, such as screws or bolts, and avoid irreversible methods (e.g., adhesives, soldering). This makes it easier to take apart the BIPV system for maintenance and upgrade, without damaging components.

Labeling and Documentation: Develop a clear labeling system and detailed documentation for assembly and disassembly. The documentation should also include a plan for the end-of-life phase of the BIPV components. The information should be easily accessible to maintenance crews and future users.

Standardisation: When possible, adopt industry-standard sizes and connections, increasing the chances of finding replacement parts and reducing the need for custom manufacturing.

Adaptability and Upgradability: Design the system with the capability to adapt to changing needs (e.g., integration of stationary electrical storage) and improvements over time.

Maintenance and Inspection: Design the system with regular maintenance and inspection in mind. Components that require more frequent replacement can be designed for easy access and replacement.

Circular Business Model (for third-party ownership): Consider implementing a circular business model to retain ownership of the BIPV system, ensuring proper disassembly, maintenance, and potential reuse. For more information, see BIPV-as-a-service business model.

Table 4.2 Environmental impact KPIs associated with the generation of 1 kWh of solar electricity, for an average grid-connected residential PV system [14].

KPI	Crystalline Silicon PV system		Thin-film PV system	
	m-Si	mc-Si	CdTe	CI(G)S
GHG emissions ($kg\ CO_2$ eq.)	42.9×10^{-3}	44.0×10^{-3}	25.5×10^{-3}	35.4×10^{-3}
Depletion of fossil fuel (MJ)	0.51	0.52	0.35	0.51
Depletion of minerals (kg Sb eq.)	5.21×10^{-6}	5.30×10^{-6}	5.23×10^{-6}	4.64×10^{-6}
Particulate matter emissions (disease incidences per million)	3.85	3.88	0.94	1.19
Acidification (kg-mol H+ eq.)	0.36×10^{-3}	0.37×10^{-3}	0.18×10^{-3}	0.21×10^{-3}
Water use (m^3)	4.49×10^{-3}	3.90×10^{-3}	0.21×10^{-3}	3.13×10^{-3}

4.8.4 Life cycle cost analysis

The investment decision for BIPV in buildings requires a complex analysis due to the involvement of various stakeholders, its impacts on the building, and the technology itself. Decision-makers such as property owners, builders, and architects always demand an assurance that is mostly determined through the project value or return. On the other hand, distributors, suppliers, and consultants tend to stress economic viability to convince their clients. The decision-makers are heterogeneous, and they hold diversified preferences with respect to economics, energy, environmental, and social factors that are influenced by many dynamics for the BIPV adoption. Thus, a preliminary life cycle cost analysis (LCCA) is particularly helpful to evaluate the overall economic feasibility and value proposition of BIPV into new or existing building projects.

The economic value of BIPV can be measured using various financial KPIs. Traditionally, BIPV has been valued at capital costs that include modules, balance-of-system (BOS), system design, permitting, construction, and labour. However, the capital cost does not reflect true BIPV value, as several other costs and benefits are experienced during the operational stages of the building. Therefore, it is important to consider the life cycle costs of a BIPV system. Levelised cost of energy, net present value, internal rate of return, and payback period are the most common life cycle metrics that have been used in practice [16,17].

Levelised cost of energy (LCOE) is an assessment of the whole life-cycle-cost over life-cycle-energy generation (e.g., €/kWh). This value is generally compared with the retail electricity price:

$$LCOE = \frac{CC + LMC + IRC - RE - SV - MRV}{LEG} \tag{4.7}$$

where CC represents capital cost, LMC represents life cycle maintenance cost, IRC represents life cycle inverter replacement cost, SV represents salvage value, MRV represents material replacement value, and LEG represents life cycle energy generation.

Net Present Value (NPV) expresses the difference between the present value of benefits and costs over the project life, normalised per kW of installed capacity to allow comparison between system alternatives (e.g., €/kW):

$$NPV = \frac{-CC - LMC - IRC + RE + MRV + SV + EBS + GS}{SC} \tag{4.8}$$

where EBS represents life cycle energy bill saving benefit, GS represents life cycle income exported to the grid, and SC represents the installed capacity of the BIPV system.

Internal rate of return (IRR) is the rate that sets NPV of the cash flows equal to zero. In other words, IRR is the interest rate (%) at which the capital investment cost is equal to the present value of future cash inflows generated by the BIPV system. IRR should not be used to evaluate investments in which further investment-after-return is required as it yields incorrect results.

Payback Period (PP) is the time required for the system to "pay back" its cost through energy cost savings or other financial benefits, and it is typically expressed in number of years.

$$PP = \frac{CC}{-LMC - IRC + RE + MRV + SV + EBS + GS} \tag{4.9}$$

As can be observed through these life cycle metrics, the standard life cycle cost parameters of a BIPV system are capital cost, maintenance and operational cost, inverter replacement cost, and residual value. The most common life cycle benefits are the economic profit of self-consumption valued at electricity price and the income of selling electricity to the grid. Despite these common parameters, indirect costs such as building material replacement, social and environmental costs, which can include power loss in transmission lines, power delivery cost, societal cost of carbon [18], and impact of BIPV on building, heating, cooling, and lighting loads (i.e., applicable for BIPV windows, skylights, and solar shading devices) have been taken into consideration to quantify the real value of BIPV. These additional benefits could uplift the economic value of BIPV favourably. Unfortunately, some costs and benefits of BIPV installations such as "corporate green status" are difficult to quantify or monetize. It is important to add all direct and indirect profits and expenses to the economic evaluations. Furthermore, decisions should not rely on a single parameter but a combination of parameters. This offers a complete valuation of BIPV from multiple perspectives based on the decision-makers' preferences and business opportunities. The following section contains a summary of BIPV-specific business models, many of which also apply to rooftop PV.

4.9 Business models

From a business point of view, BIPV systems can be envisioned as a long-term investment. To estimate the system's cash flow over time and assess life cycle costs through metrics such as LCOE, NPV, and PP, factors such as electricity generation value, incentives, installation and maintenance costs, taxes, and debt should be considered. In terms of electricity generation value, there are various models for grid-connected BIPV systems, with the most common ones outlined as follows [19]. These models provide opportunities for accessing solar energy with or without the upfront costs of system ownership, but they have different ownership structures or financial arrangements that should be carefully evaluated. Ultimately, the choice depends on individual preferences, financial goals and resources, and the local electricity market.

4.9.1 Feed-in tariff

Feed-in tariff (FiT) refers to the incentive policies that aim to promote investment in photovoltaic systems and accelerate their market adoption. Although market variations exist, FiTs usually involve long-term contracts (10 to 25 years) between the local utilities and the building owner (assuming that is the same as the BIPV system owner). The contract guarantees grid connection for the contract period and creates the obligation for the utility to purchase any kilowatt-hour (kWh) of excess generated solar electricity, at a premium above or additional to the retail price. FiT programs often incorporate "tariff degression," a mechanism where the utility purchase price gradually decreases over time until it matches retail rates. Ultimately, FiTs require a capital investment for the purchase and installation of the BIPV system and, in some cases, it might be eligible for tax credits or deductions. FiT programs aim to provide building owners with cost-based compensation, offering price stability and making it faster to recoup the capital investment. FiTs are typically implemented in emerging PV markets, aimed at encouraging (BI) PV adoption. In more mature PV markets, FiTs are often replaced by net-metering programs.

4.9.2 Net-metering

Net-metering allows building owners to sell excess generated solar electricity to the grid as energy credits. Depending on the utility or power regulator, the kilowatt-hour (kWh) value may follow a fixed rate, time-of-use (ToU), or real-time dynamic pricing based on demand response, using a bi-directional smart meter. Under ToU, prices fluctuate based on predetermined demand periods, rising during peak hours, and dropping during off-peak hours like overnight. The ToU rates change semi-annually between cooling and heating seasons. Real-time pricing can shift as often as hourly, reflecting electricity generation or purchase costs at the grid level. Less often, excess solar electricity is credited at a rate below retail rates to incentivise self-consumption and discourage grid export.

In a net-metering arrangement, building owners do not receive a check from exporting electricity to the grid. Instead, they receive energy credits that can be used to "offset" money-wise their future consumption (e.g., up to 36 months from the billing time) and lower their electricity bills, effectively using the power grid as electrical storage. Like FiT, net-metering requires a capital investment for the purchase and installation of the BIPV system and has a longer payback period (PP). In addition, net-metering is usually eligible for financial subsidies such as tax credits, deductions, and accelerated depreciation incentives.

4.9.3 Power purchase agreement

A power purchase agreement (PPA) is a contract between a third-party solar energy provider (e.g., a solar developer or installer) and an electricity consumer (e.g., building owner or property managers) that enables the consumer to purchase electricity generated by the BI(PV) system without having to own or maintain the system itself. Although uncommon in BIPV projects, PPAs are more prevalent in large PV rooftop installations (100–250 kW). Consumers pay for the electricity generated by the system at a predetermined rate per kilowatt-hour (kWh) for the period of the agreement (10 to 25 years), typically at a lower rate than retail price. A kilowatt-hour rate escalator is common, accounting for the system's decreasing efficiency, rising operational and maintenance costs, and increasing electricity retail rates.

In the case of a PPA, the consumer is not eligible for any renewable tax credits or incentives as the solar provider retains ownership, operation, and maintenance duties, requiring no capital investment from the customer. Exiting a PPA before the termination date may come with financial penalties or requirements to buy out the contract. At the end of the agreement, the building owner may extend the agreement, purchase the BI(PV) system, or have it uninstalled from the building.

4.9.4 BIPV-as-a-service

BIPV-as-a-service, or BIPV lease, is a leasing contract between a third-party solar provider and the consumer. Like a PPA, the provider owns and maintains the BIPV system for the lease duration. However, in the case of a BIPV lease, consumers pay a set monthly fee, typically spanning for 15 years, regardless of the energy they use from the BIPV. This fee often includes an annual escalator for inflation and system depreciation.

In a BIPV lease, consumers do not get tax credits or incentives. This is because the solar provider retains system ownership and maintenance, eliminating the need for a capital investment from the consumer. Depending on the lease terms, consumers might buy, renew, or return the BIPV system, at the end of the contract. Some contracts also offer system upgrades or early renewal options.

Note

1. One sun-hour is equal to 1 kwh/m^2 per day. Multiply sun-hours × 365 to convert to kwh/m^2/year.

References

[1] CanmetENERGY, "Planning and decision guide for solar PV systems," *Natural Resources Canada*, M154-135/2020E-PDF.

[2] L. Lisell, T. Tetreault, and A. Watson, "Solar ready buildings planning guide," *NREL/TP-7A2-46078*, 970752, Dec. 2009, doi:10.2172/970752.

[3] F. Garde, D. Aelenei, L. Aelenei, A. Scognamiglio, and J. Ayoub, *Solution Sets for Net-Zero Energy Buildings: Feedback from 30 Buildings Worldwide*. Wiley-VCH Verlag GmbH & Co. KGaA, 2017. doi:10.1002/9783433604663.

[4] P. Macé, D. Larsson, and J. Benson, "Inventory on existing business models, opportunities and issues for BIPV," *International Energy Agency (IEA) PVPS Task 15, T15-03*, 2018, https://iea-pvps.org/wp-content/uploads/2020/02/Inventory_on_Existing_Business_Models_Opportunities_and_Issues_for_BIPV_by_Task_15.pdf.

[5] F. Bre, R. M. E. Silva Machado, L. K. Lawrie, D. B. Crawley, and R. Lamberts, "Assessment of solar radiation data quality in typical meteorological years and its influence on the building performance simulation," *Energy Build.*, vol. 250, p. 111251, Nov. 2021, doi:10.1016/j.enbuild.2021.111251.

[6] N. Martín-Chivelet and D. Montero-Gómez, "Optimizing photovoltaic self-consumption in office buildings," *Energy Build.*, vol. 150, pp. 71–80, Sep. 2017, doi:10.1016/j.enbuild.2017.05.073.

[7] I. Sartori, A. Napolitano, and K. Voss, "Net zero energy buildings: A consistent definition framework," *Energy Build.*, vol. 48, pp. 220–232, May 2012, doi:10.1016/j.enbuild.2012.01.032.

[8] N. Jakica, R. J. Yang, and J. Eisenlohr, "BIPV design and performance modelling: Tools and methods," *International Energy Agency (IEA) PVPS Task 15, T15-09*, IEA PVPS, 2019, https://iea-pvps.org/wp-content/uploads/2020/01/IEA-PVPS_15_R09_BIPV_Design_Tools_report.pdf.

[9] D. L. Evans, "Simplified method for predicting photovoltaic array output," *Sol. Energy*, vol. 27, no. 6, pp. 555–560, 1981, doi:10.1016/0038-092X(81)90051-7.

[10] W. De Soto, S. A. Klein, and W. A. Beckman, "Improvement and validation of a model for photovoltaic array performance," *Sol. Energy*, vol. 80, no. 1, pp. 78–88, Jan. 2006, doi:10.1016/j.solener.2005.06.010.

[11] D. King, W. Boyson, and J. Kratochvil, "Photovoltaic array performance model," *SAND2004-3535*, 919131, 2004, doi:10.2172/919131.

[12] J. Lopez-Lorente, et al., "Characterizing soiling losses for photovoltaic systems in dry climates: A case study in Cyprus," *Sol. Energy*, vol. 255, pp. 243–256, May 2023, doi:10.1016/j.solener.2023.03.034.

[13] R.E. Pawluk, Y. Chen, and Y. She. Photovoltaic electricity generation loss due to snow–A literature review on influence factors, estimation, and mitigation. Renewable and Sustainable Energy Reviews, no. 107, pp 171–182, 2019, doi:10.1016/j.rser.2018.12.031.

[14] PACTE, "Systèmes photovoltaïques par modules rigides intégrés en couverture," *Le Programme d'Action pour la qualité de la Construction et la Transition Energétique, DGALN/DHUP-978-2-11-151809-4*, PACTE, 2017, https://www.proreno.fr/documents/systemes-photovoltaiques-par-modules-rigides-integres-en-couverture.

[15] R. Frischknecht, P. Stolz, L. Krebs, M. de WildScholten, and P. Sinha, "Life cycle inventories and life cycle assessment of photovoltaic systems," *International Energy Agency (IEA) PVPS Task 12, T12–19*, IEA PVPS, 2020, https://iea-pvps.org/wp-content/uploads/2020/12/IEA-PVPS-LCI-report-2020.pdf.

[16] R. P. N. P. Weerasinghe, et al., "Economic viability of building integrated photovoltaics: A review of forty-five (45) non-domestic buildings in twelve (12) western countries," *Renew. Sustain. Energy Rev.*, vol. 137, p. 110622, Mar. 2021, doi:10.1016/j.rser.2020.110622.

[17] W. Short, D. J. Packey, and T. Holt, "A manual for the economic evaluation of energy efficiency and renewable energy technologies," *National Renewable Energy Lab. (NREL)*, Golden, CO (United States), NREL/TP-462–5173, Mar. 1995, doi:10.2172/35391.

[18] H. Gholami and H. N. Røstvik, "Economic analysis of BIPV systems as a building envelope material for building skins in Europe," *Energy*, vol. 204, p. 117931, Aug. 2020, doi:10.1016/j.energy.2020.117931.

[19] A. Campoccia, L. Dusonchet, E. Telaretti, and G. Zizzo, "An analysis of feed'in tariffs for solar PV in six representative countries of the European Union," *Sol. Energy*, vol. 107, pp. 530–542, Sep. 2014, doi:10.1016/j.solener.2014.05.047.

CHAPTER 5

Design of BIPV envelope and case studies

Costa Kapsis, Ana Marcos Castro, and Nuria Martín Chivelet

5.1 Introduction

From an engineering and standardisation perspective, BIPV applications can be categorised into roofs, façades, and externally integrated systems [1]. Each application category features a common set of performance requirements that determine the design and construction needs of the BIPV systems and products.

Chapter 5 aims to consolidate existing BIPV design and construction knowledge and produce technical reference drawings for various BIPV applications [2,3]. The engineering solutions aim to offer inspiration on how BIPV systems could be designed and constructed, and they do not constitute exact solutions. The responsibility lies with the building professionals to ensure that the final design and construction fulfill all project requirements, always in compliance with the local building, electrical, and fire codes and regulations. Annotated reference drawings are accompanied by technical guidelines and "mistakes to avoid." These resources cater to various climates and building functions, applicable to both new constructions and retrofits. The aim is to equip building professionals with technical insights to craft distinctive solutions that harmonise performance and aesthetics.

The chapter presents reference drawings for BIPV roof systems, façade applications, and solar shading devices followed by photovoltaic applications that stretch beyond buildings to infrastructure in the built environment (e.g., parking canopies and transit shelters). Each application section is followed by a set of high-profile BIPV case studies from around the globe. The list covers a wide range of climates, building uses (e.g., commercial, residential, convention centres and arenas), and applications.

Each case study includes a building description, BIPV system details with key characteristics, and a technical drawing for insights into design, installation, and operation. A "quick overview" with key stakeholders (e.g., architect, product manufacturer, installer), local climate details (climate classification, annual solar irradiation), and system performance metrics (BIPV array nominal power, annual electricity generation per BIPV module area, and annual electricity generation) is also provided. Annual electricity generation for BIPV projects built before 2022 is averaged from several years of data, whereas for projects from 2022 onwards, it's estimated through climate-based energy simulations. Additionally, each case study is accompanied by a heatmap displaying the annual PV potential generation for various orientations and tilt angles, aiding future BIPV designers. Note that the case studies of Deltarosso, Azurmendi restaurant, and Solaris 416 feature multiple BIPV applications but are presented under one section, showcasing the primary application of a BIPV façade, skylight, and rainscreen, respectively.

5.1.1 PV potential heatmap

A simple way to quickly assess the potential electricity generation of a BIPV system based on its orientation and tilt angle is by using a potential heatmap graph. Each concentric ring represents the tilt angle of the BIPV system (0° for horizontal to 90° for vertical installations) while each spoke represents the orientation of the system.

For instance, using Figure 5.1, the building designer can quickly estimate that for the city of Madrid, an unshaded BIPV

DOI: 10.1201/9781003432241-5

roof that faces south (S) and is installed at a tilt angle of 30° is estimated to generate 1,400–1,600 kWh/kW. This means that if the proposed BIPV skylight is 5 kW, then the estimated annual electricity generation is 7,000–8,000 kWh/year. Similarly, an unshaded 20 kW BIPV vertical rainscreen that faces southwest (SW) is estimated to generate 20,000–24,000 kWh/year (i.e., 20 kW × [1,000, 1,200] kWh/kW = [20,000, 24,000] kWh).

Note that the heatmap provides a quick estimation that can be used at the preliminary design stage of a BIPV project. For a more accurate estimation including shading losses, it is advisable to conduct detailed BIPV design and optimisation during the design development phase using either a photovoltaic simulation software or building performance simulations. The heatmaps were generated using *Meteonorm* weather data as an input for each city. The analysis assumed a fully unobstructed BIPV system, using a crystalline Silicon module with efficiency of 21%, temperature coefficient of power of 0.35%/°C, inverter efficiency of 96%, system losses of 14%, and DC to AC size ratio of 1.2.

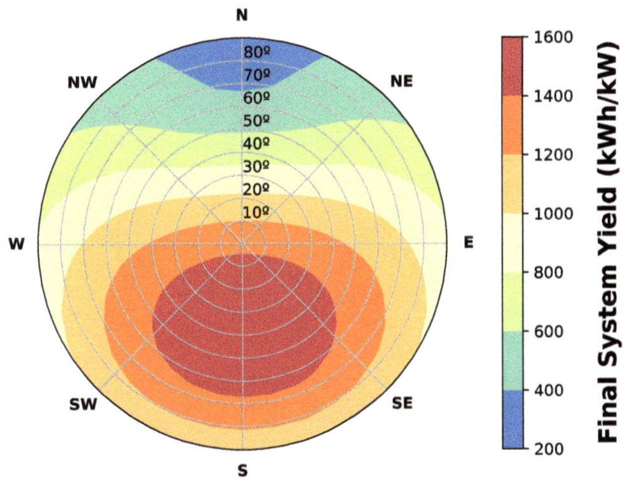

Figure 5.1 PV potential heatmap for Madrid: Annual PV energy generation potential for an unshaded BIPV system in Madrid. Credits: Ana Marcos Castro, CIEMAT.

5.1.2 Case studies

This chapter includes detailed description of 24 case studies covering different BIPV applications and countries. Table 5.1 lists the case studies in order of appearance in the chapter.

Table 5.1 List of the case studies described in the chapter.

Case study	Country	BIPV application	Construction
Varennes Library	Canada	Roof	New
Umwelt Arena	Switzerland	Roof	New
Folzano Kindergarten	Italy	Roof	New
Solaris 416	Switzerland	Roof and rainscreen	New
Glasberga Residential District	Sweden	Roof	New
Andreas Bjørns St.1	Denmark	Roof	Retrofit
Urban Park In Certosa Island	Italy	Roof	Retrofit
San Antón Market	Spain	Skylight	Retrofit
Azurmendi Restaurant	Spain	Skylight and curtain wall	New
Edmonton Convention Centre	Canada	Skylight	Retrofit
Porta Susa Av Station	Italy	Skylight	New
Jeanne & Peter Lougheed Performing Arts Centre	Canada	Rainscreen	New
Tuas Port Administrative Building	Singapore	Rainscreen	New
DeltaROSSO	Switzerland	Rainscreen	New
Manitou A Bi Bii Daziigae University Building	Canada	Rainscreen	New
Fanshawe College Innovation Village	Canada	Rainscreen	New
Genyo Research Facility	Spain	Double-skin façade	New
Palazzo Argonauta	Italy	Rainscreen	Retrofit
Yanmar Tokyo Building	Japan	Curtain wall	New
NCC Headquarters	Sweden	Curtain wall	New
Palazzo Lombardia	Italy	Curtain wall	New
Gothenburg Garage	Sweden	Solar shading	New
Franklin University Student Residence	Switzerland	Solar shading	New
Arrival Center in Schönbrunn	Austria	Carport canopy	New

5.2 BIPV roof systems

5.2.1 Discontinuous roof

General: A BIPV roof system is one of the most common BIPV building applications. It can effectively replace an existing shingle roof on a heritage building or be the roofing system for a new construction [4]. Its design principles closely adhere to traditional roofing systems with the inclusion of electrical requirements and structural loading.

Structural Design: The roof structure ① should be designed to withstand both the static load of the BIPV roof's weight and the environmental loads such as wind, snow, and ice (when applicable). When local building codes do not provide design guidelines, the additional design dead load of at least 0.2 kN/m^2 can be used to accommodate for the weight of glass-glass modules and the mounting hardware of the BIPV roof system [5]. To mitigate thermal expansion of the modules, a BIPV mounting system using a "shingle overlap" approach can be installed, where each row of modules minimally overlaps with the row beneath it.

Hygrothermal Performance: The BIPV modules or shingles ⑫ play the role of the weather barrier, deflecting most of the direct rain and providing protection from sun, snow, and ice. A naturally ventilated cavity ⑨ provides adequate airflow for the BIPV modules, prevents overheating, and lets any penetrating moisture drain toward the gutter ⑯ or evaporate and vent to the outside through the roof ridge ⑰. For this reason, the air cavity thickness between the roof sheathing and the BIPV modules should be 100 mm or more [6]. In snowy regions, the air cavity can also reduce the formation of "ice dams" at the roof edge. An insect screen ⑮ at the top and bottom course of the air cavity is advised. An electrical conduit ⑱ with a weather head should be used to bring the DC wiring to the electrical room. Depending on the sheathing product, the application of a primary water barrier ⑥ might be required (e.g., adhered membrane or spray applied) on the exterior face of the roof sheathing to eliminate any further water ingress and ensure a dry roof assembly. The choice of rigid thermal insulation ④ material and thickness depends on local building codes and sustainability requirements. An alternative assembly using flexible a-Si thin-film modules can be installed using corrugated metal sheets. For more information, see the Folzano Kindergarten case study.

Fire Safety: It may be necessary to apply thermal and ignition barriers to the interior surface of the unvented roof assembly and structure. Some fire codes may also require the roof sheathing ⑤ (including the fascia and the soffit) to be fire-resistant. It is also a good practice for areas prone to wildfires, as embers can enter the vented cavity and potentially start a roof fire. A rapid shutdown switch is required, easily accessible for the fire brigade in the event of fire.

Electrical Considerations: Effective cable management and labelling ensure BIPV roof longevity. The use of cable trays ⑩, clips, and ties can help secure the wiring, prevent mechanical stress, and avoid abrasion from sharp corners and movement due to wind or other vibrations that could damage wiring or connectors. The connectors should have a minimum rating of IP 65, while the DC cables should be installed to avoid induction loops that can generate significant magnetic fields. When possible, the modules should be connected in a leapfrog fashion (vs daisy chain wiring). For complex roof layouts, the use of microinverters or optimizers is recommended as they allow for less uniform panel placement and can adapt to challenging shading conditions. For locations where substantial shading is of concern (e.g., roof valleys), the BIPV modules can be replaced with identical-looking dummy modules, creating a uniform roof finish.

Architectural Considerations: BIPV roofing products come in various colours, textures, and shapes, seamlessly blending with the architectural style of a building (e.g., from terracotta-coloured PV tiles to flexible a-Si thin-film shingles). When it comes to a roof layout, a gable or a shed roof style is recommended to maximise the available solar surface and for ease of installation. At the design development stage of the project, the BIPV roof dimensions can be adjusted to accommodate a precise number of the selected BIPV roof products. Obstructions such as vents, chimneys, and architectural features should be located north of the photovoltaic area (south, for the southern hemisphere) to avoid shadow casting, when possible. In snow-prone areas, the use of frameless BIPV products installed at a roof slope of 30 degrees or more will significantly reduce snow retention and increase the energy yield of the system over winter months. In these cases, the use of snow guards is necessary and, thus, the last one or two rows of BIPV modules or shingles located at the roof edge should be dummy modules ⑭.

Sustainability: The BIPV roof should be engineered to last for at least 35 years. If the building is in warm or hot regions, the use of light-coloured BIPV roofing products can assist with the mitigation of the urban heat island effect. However, light-coloured BIPV products tend to have lower electrical efficiencies than conventional PV modules, resulting in longer energy payback time. In areas where energy resiliency is of importance (e.g., areas prone to floods, earthquakes, hurricanes, heat waves, etc.), the use of grid-interactive inverters coupled with stationary electric storage is recommended.

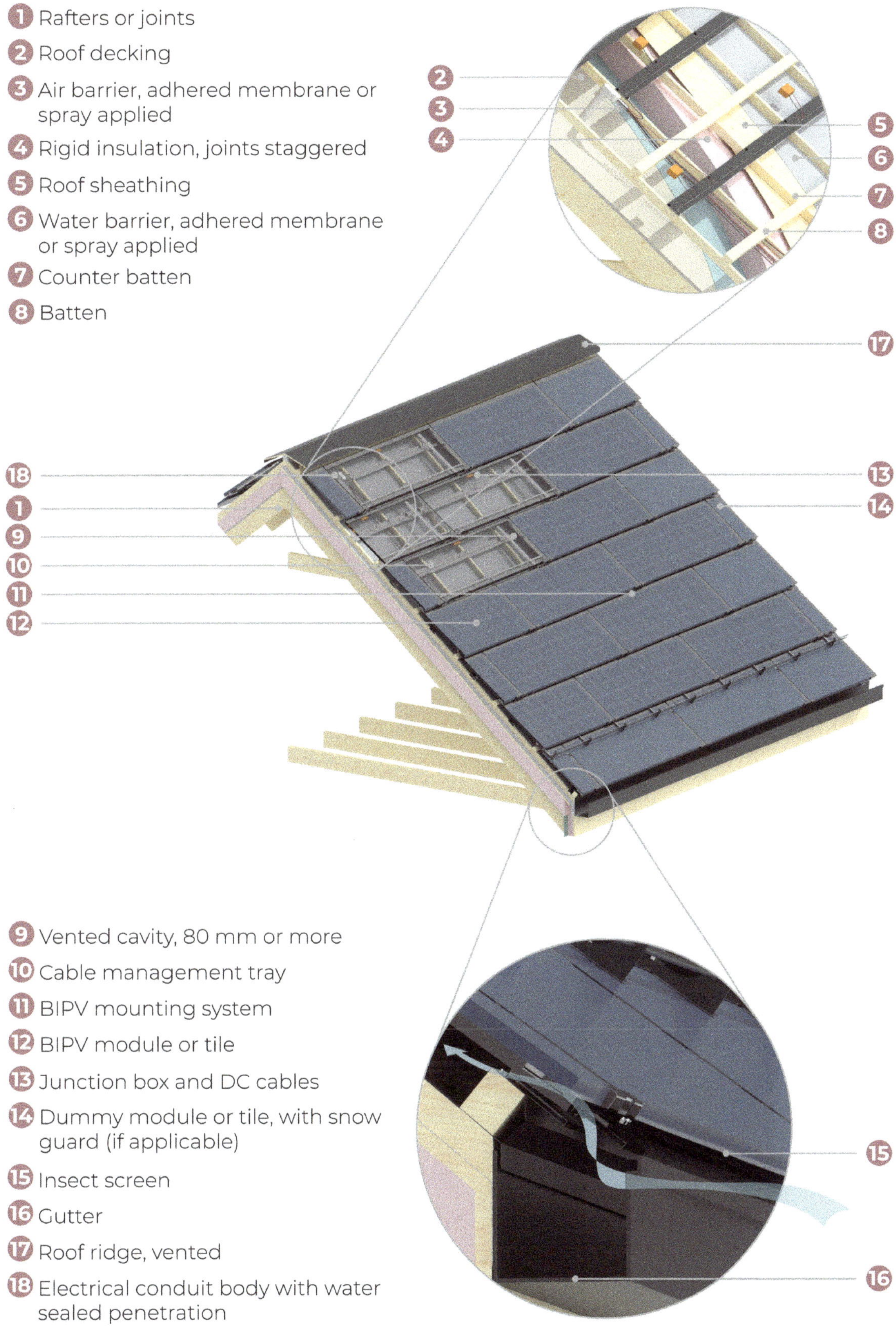

Figure 5.2 Axonometric representation of a typical BIPV roof system. Credits: Sara Eskandar, Ana Marcos Castro, Marley Dowling, Patrick Angkiriwang.

Varennes library

LOCATION	**Varennes, Canada (45° 40′ 56″ N)**
USE	**Library**
PV PRODUCER	**Heliene**
ARCHITECT	**Labbé, Laroche/Gagné, Leclerc et Associés**
ENGINEER	**Stantec**
INSTALLER	**Bordeaux Électrique Inc**
YEAR OF CONSTRUCTION	**2014**

Project description: The Varennes library is a net-zero building in Varennes, Quebec, covering a total area of 2,230 m². Built to replace the former library, this LEED Gold certified building was inaugurated in 2016 and is spread across two floors. The BIPV installation is composed of 428 PV modules, with a total surface area of 711 m².

BIPV system key characteristics: The south-oriented BIPV roof of the building is split into three sections: naturally ventilated, mechanically ventilated, and mechanically ventilated with heat recovery. During the heating season, the pre-heated fresh air from the heat recovery section of the BIPV roof is introduced into the air handling unit (AHU) to supplement the fresh air supply of the library and reduce the energy demand for heating. The modules used are all black (i.e., black cells/backsheet/frame) for minimal aesthetics while potentially reducing snow retention.

1.4 MWh/m²
Annual horizontal solar irradiation

Humid continental – Dfb
Köppen climate classification

Roof
BIPV application

New
Type of construction

Opaque m-Si
PV technology

110 kW
BIPV array nominal power

169 kWh/m²
Annual electricity generation per BIPV module area

120 MWh
Annual electricity generation

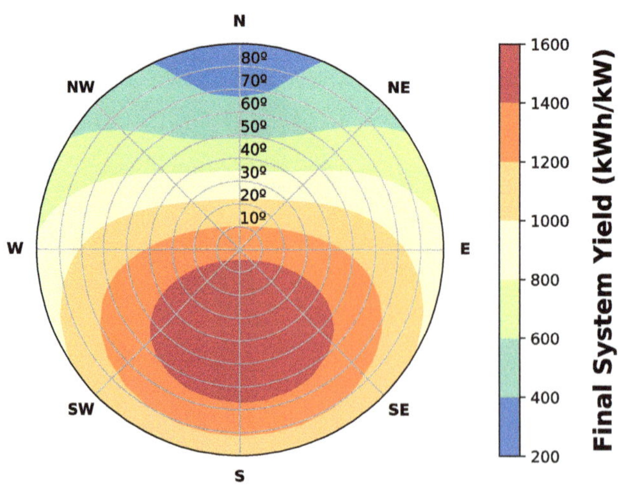

Figure 5.7 PV potential heatmap for Varennes, Canada. Credits: Ana Marcos Castro, CIEMAT.

Figure 5.3 Varennes Public Library. Credits: Stantec Inc.

Figure 5.4 Varennes library under construction. Credits: Jose Candanedo.

Figure 5.5 Panoramic view of Varennes library. Credits: Marie D Martel, Wikimedia Commons.

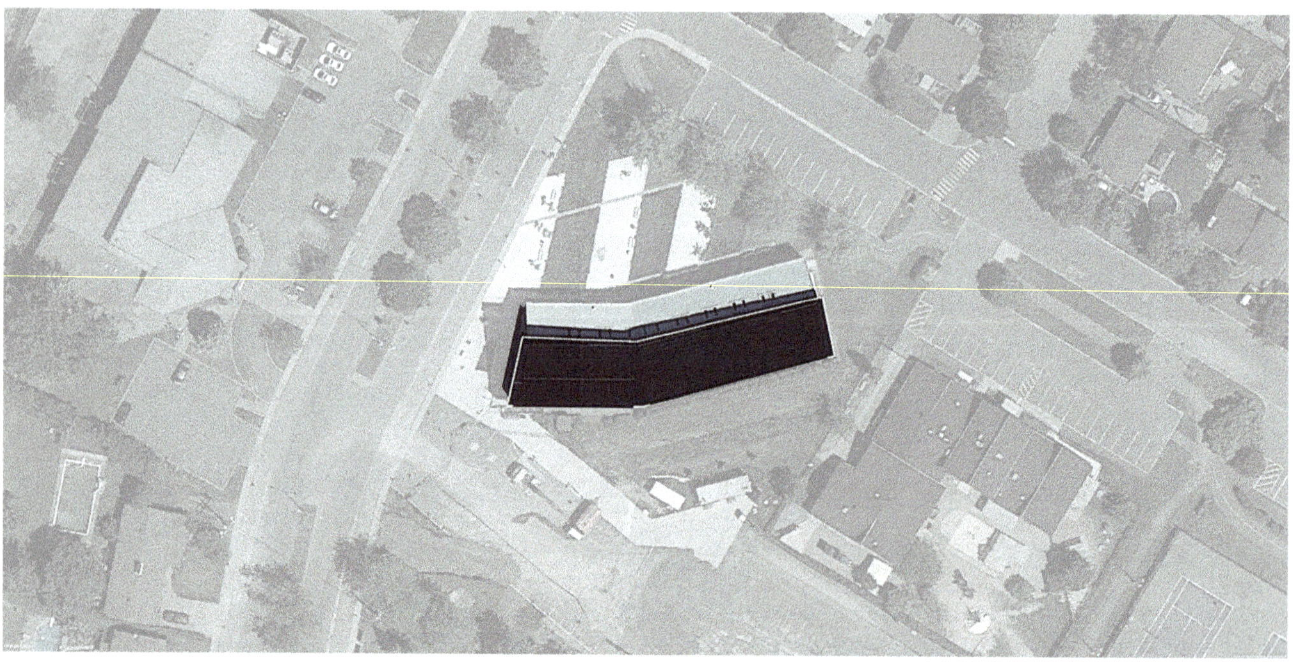

Figure 5.6 Site plan of Varennes Library. Source: Google Maps.

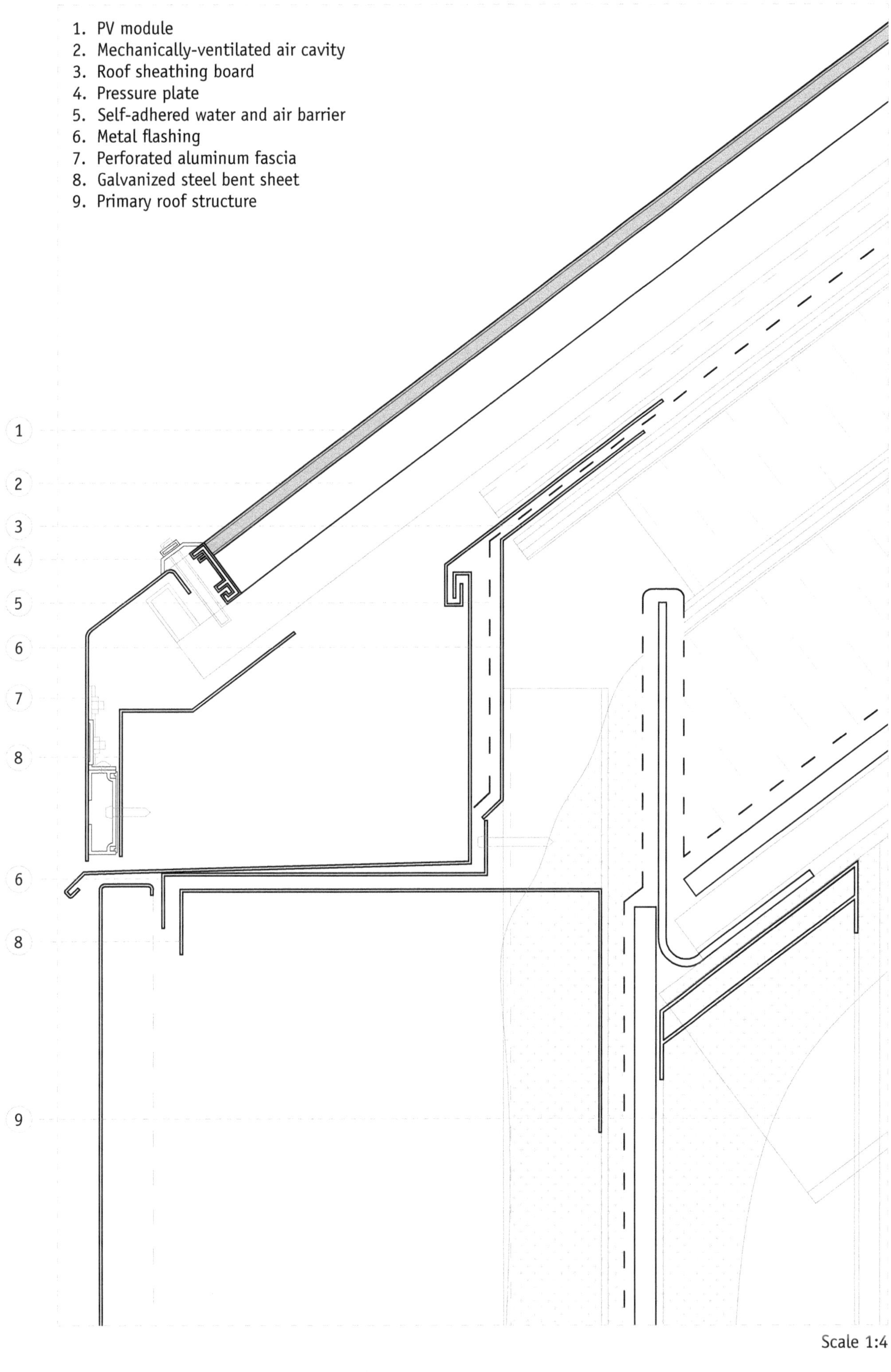

Figure 5.8 Varennes Library. Cross-section drawing of the BIPV roof installation. Credits: adapted by Ana Marcos Castro from Labbé, Laroche/Gagné, Leclerc ©.

Umwelt Arena

LOCATION	Spreitenbach, Switzerland (47°25′29″ N)
USE	Exhibition centre
CLIENT	Umwelt Arena AG Spreitenbach
PV PRODUCER	3S Swiss Solar Systems
ARCHITECT	René Schmid Architekten
INSTALLER	BE Netz AG
YEAR OF CONSTRUCTION	2012

Project description: The Umwelt Arena Schweiz (Swiss Environmental Arena) in Spreitbach is an exhibition centre and a technology hub that promotes sustainability, ecology, and energy efficiency. Throughout the year, the building hosts several seminars, trade fairs, congresses, and events. A central atrium with a racetrack is used for indoor test drives on e-bikes, e-scooters, and e-cars. With a 12,700 m² of net floor area and 20-meter height, the large arena is covered with a 5,333 m² BIPV roof. The building uses an innovative heating, cooling, and ventilation system running on solar power through a thermally activated building structure (TABS) in the concrete floors and connected to a geo-exchange system located below the underground parking.

BIPV system key characteristics: The large BIPV roof has a unique octagonal shape, resembling a tension tent, that consists of 33 triangular surfaces of various orientations and slopes. It integrates 3,663 frameless rectangular BIPV modules (1300 mm × 875 mm × 6.5 mm) and 1,644 customised ones, all made of safety glass with a black backsheet. The BIPV modules are supported by a roof construction made of insulated wooden box girder elements. The annual electricity generation of the BIPV roof covers 200% of the arena's annual energy consumption, achieving net-positive energy and carbon-neutral performance targets. In 2012, the arena received the Solar Award from the Norman Foster Foundation.

1.2 MWh/m²
Annual horizontal solar irradiation

Humid temperate – Cfb
Köppen climate classification

Roof
BIPV application

New
Construction

Opaque m- Si
PV technology

737 kW
BIPV array nominal power

101.3 kWh/m²
Annual electricity generation per BIPV module area

540 MWh
Annual electricity generation

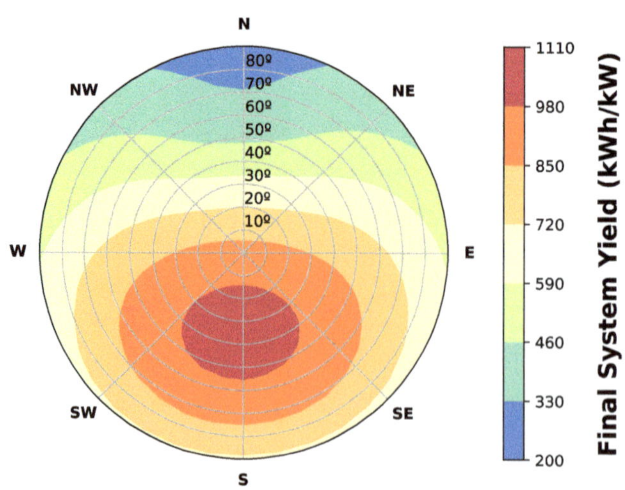

Figure 5.13 PV potential heatmap for Spreitenbach, Switzerland.

Figure 5.9 Panoramic view of Umwelt Arena. Credits: Isa Zanetti, SUPSI.

Figure 5.10 Eagle-eye view of Umwelt Arena. Credits: Isa Zanetti, SUPSI.

Figure 5.11 Close-up of the BIPV roof at Umwelt Arena. Credits: Isa Zanetti, SUPSI.

Figure 5.12 Site plan of Umwelt Arena. Source: Google Maps.

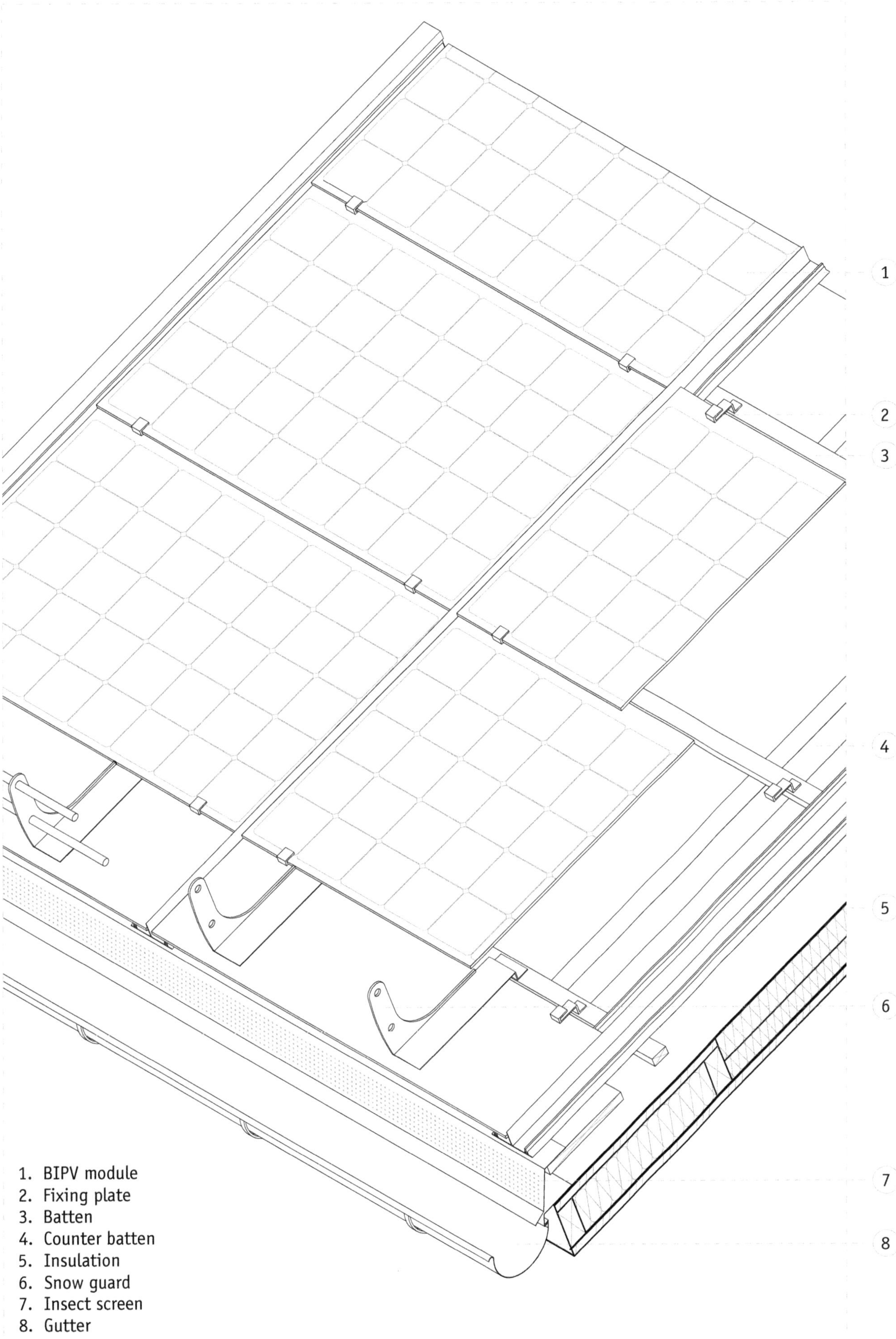

Not to scale

Figure 5.14 Umwelt Arena. Axonometric drawing of the BIPV installation. Credits: Adapted by Ana Marcos Castro from René Schmid Architekten ©.

Folzano kindergarten

LOCATION	Folzano, Italy (45° 29′ 45″ N)
USE	Kindergarten
PV PRODUCER	Kalzip® AluPlusSolar-B 68 with Uni-Solar a-Si modules
ARCHITECT	Pietrobelli e Zizioli
ENGINEER	G.Cremaschini, U. Bianchini
INSTALLER	METALCOOP
YEAR OF CONSTRUCTION	2010

Project description The kindergarten was completed in 2010, and it was designed with energy efficiency and children's wellness in mind. The bioclimatic design follows the sun path while avoiding the prevailing winds. This generates a geometry that defines the inside of the building due to the oval shape and varying height of the roof, which increases towards the outer façade to create a tilted effect.

BIPV system key characteristics: The flexible a-Si thin-film modules use aluminium as a backsheet and are 2.849 m × 0.537 m and weigh 9.2 kg. The long BIPV modules are incorporated in a south-tilted (25° to 35°) radial roof pattern of the main building and the greenhouse. The total covered BIPV surface of 168 m² provides 10.4 kW array nominal power and generates 13 MWh per year. The BIPV roofing system, which also integrates an underlying layer of insulation, does not require fixing components for the modules and ensures protection for the electric cables, which are positioned in the roof intrados, allowing for easy inspection and maintenance.

1.3 MWh/m²
Annual horizontal solar irradiation

Humid subtropical – Cfa
Köppen climate classification

Roof
BIPV application

New
Type of construction

Opaque amorphous-Si
PV technology

10.4 kW
BIPV array nominal power

77 kWh/m²
Annual electricity generation per BIPV module area

13 MWh
Annual electricity generation

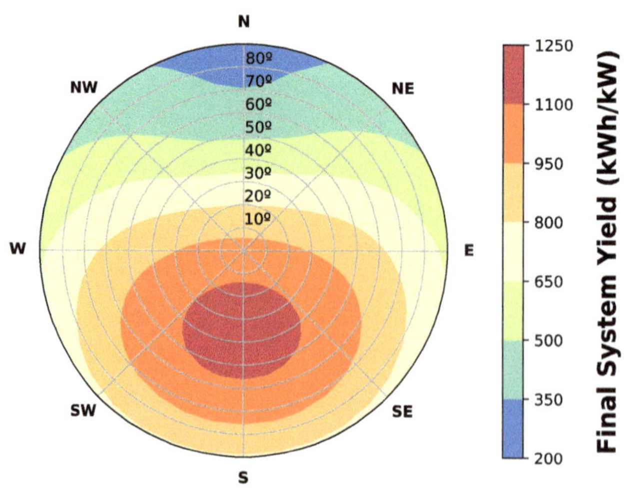

Figure 5.18 PV potential heatmap for Folzano, Italy. Credits: Ana Marcos Castro, CIEMAT.

Figure 5.15 Panoramic view of Folzano Kindergarten. Credits: Jennifer Adami, EURAC.

Figure 5.16 Close up of Folzano Kindergarten BIPV roof. Credits: Jennifer Adami, EURAC.

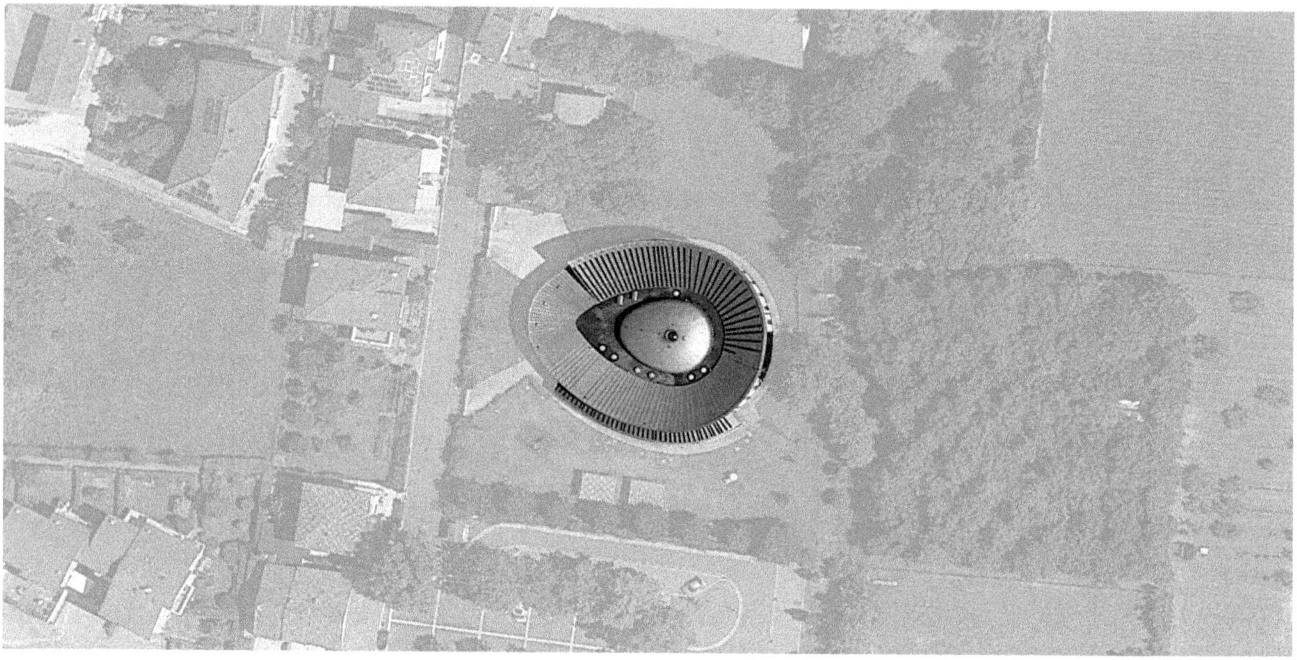

Figure 5.17 Site plan of Folzano Kindergarten. Credits: Google Maps.

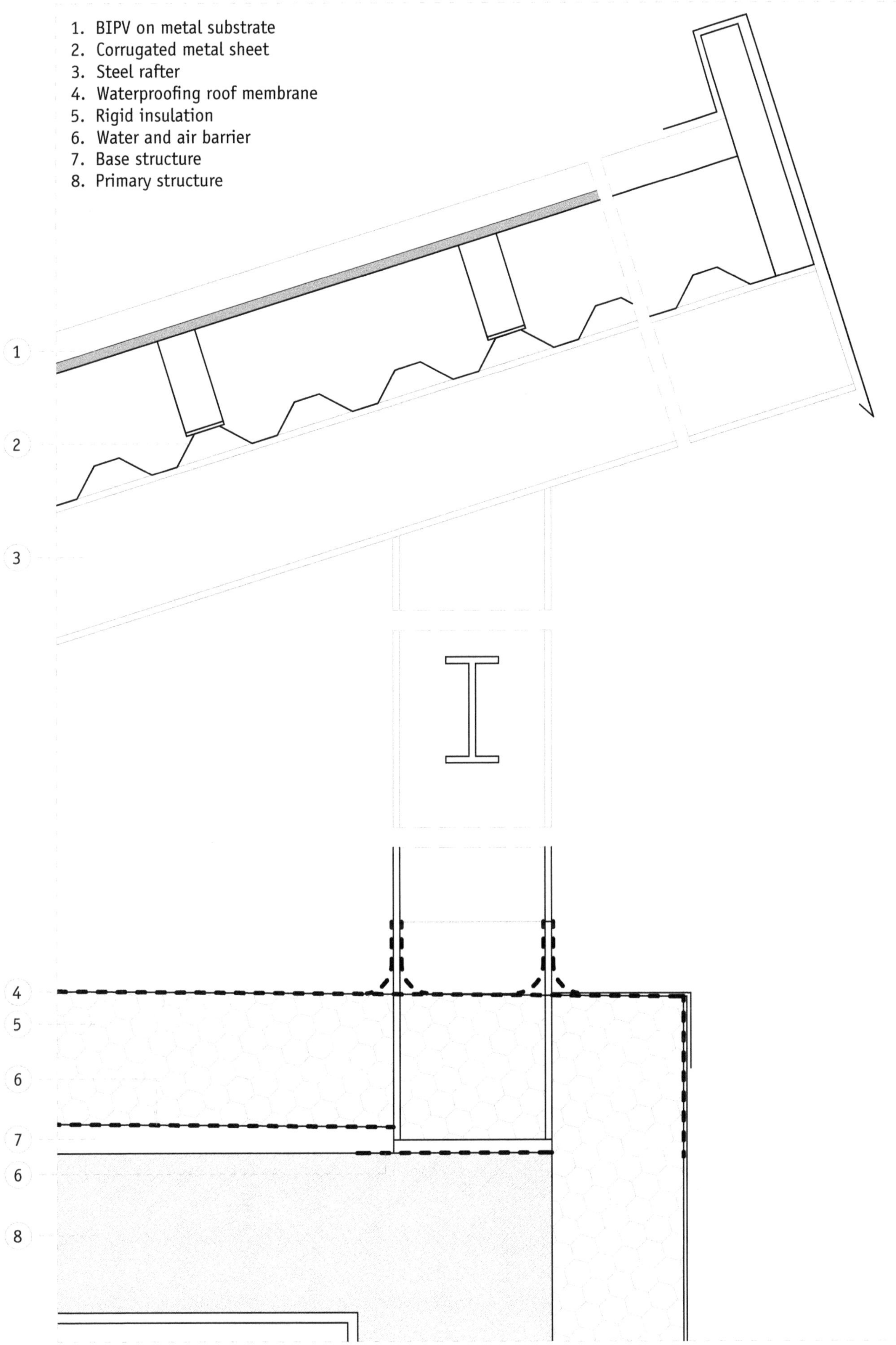

Scale 1:8

Figure 5.19 Folzano Kindergarten. Cross-section drawing of the BIPV roof installation. Credits: adapted by Ana Marcos Castro from Pietrobelli e Zizioli ©.

Solaris 416

LOCATION	**Zurich, Switzerland (47° 20′ 27″ N)**
USE	**Residential, multi-family**
PV PRODUCER	**Ertex Solartechnik GmbH**
ARCHITECT	**Huggenbergerfries Architekten AG**
INSTALLER	**Suntechnics Fabrisolar & Scherrer Metec AG**
YEAR OF CONSTRUCTION	**2017**

Project description: This multi-storey residential building hosts ten apartments with views to Lake Zurich. The building envelope includes 1300 BIPV modules from Ertex Solar that cover 400 m² of the southern and western façades and 200 m² of the roof. The building uses biogas for heating and is electrically self-sufficient thanks to the BIPV system. The excess of generated PV electricity is stored in a 10 kWh stationary battery and one electric car.

BIPV system key characteristics: The architects wanted the BIPV modules not to be recognised and chose a brown cladding made of ribbed cast glass that hides the PV cells. The Lucerne University of Applied Sciences collaborated with the modules' colour development. Depending on the different light conditions and view angles, the opaque homogeneous BIPV cladding ranges from dark red and shiny silver to violet.

1.2 MWh/m²
Annual horizontal solar irradiation

Humid temperate – Cfb
Köppen climate classification

Roof and rainscreen
BIPV application

New
Type of construction

Opaque coloured m-Si
PV technology

71 kW
BIPV array nominal power

53 kWh/m²
Annual electricity generation per BIPV module area

31.8 MWh
Annual electricity generation

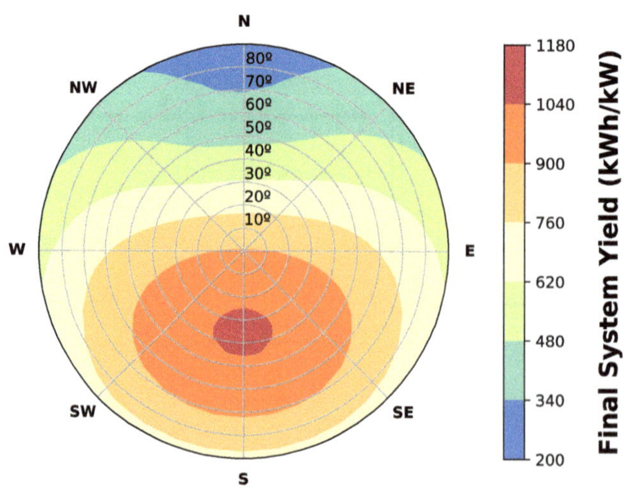

Figure 5.24 PV potential heatmap For Zurich, Switzerland. Credits: Ana Marcos Castro, CIEMAT.

Figure 5.20 Front view of the Solaris 416 building. Credits: Isa Zanetti, SUPSI.

Figure 5.21 Eagle-eye view of Solaris 416. Credits: Isa Zanetti, SUPSI.

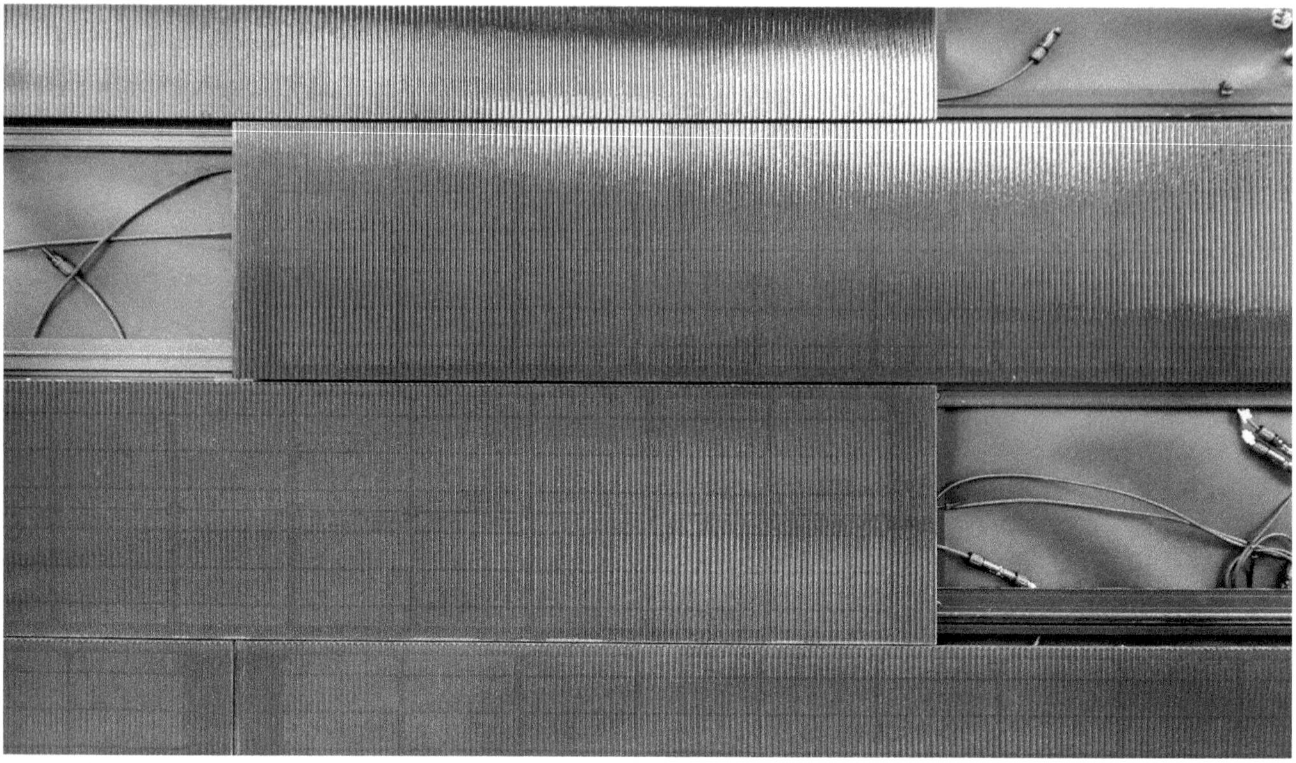

Figure 5.22 Detail of the BIPV façade at Solaris 416. Credits: Isa Zanetti, SUPSI.

Figure 5.23 Site plan of Solaris 416. Source: Google Maps.

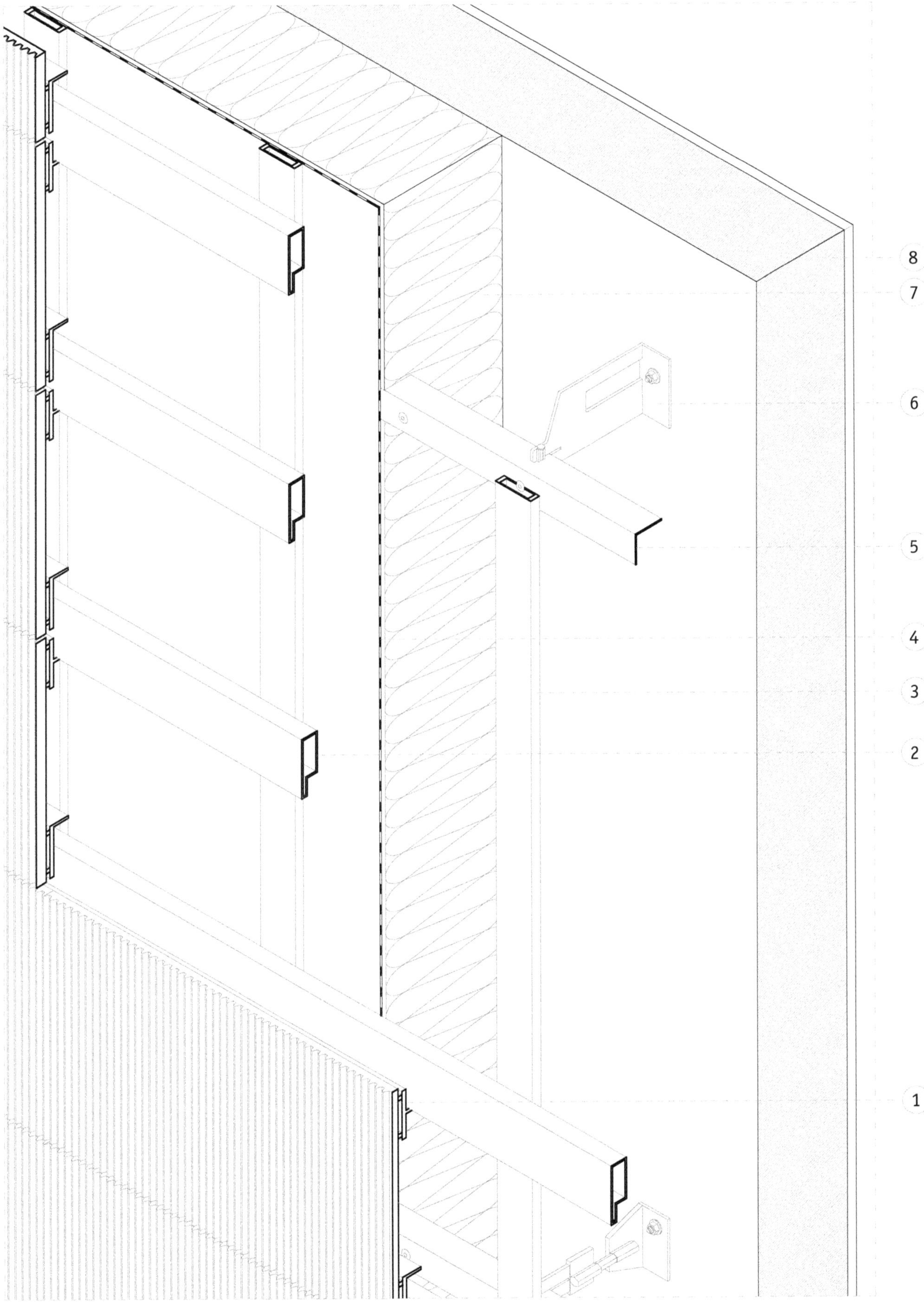

1. BIPV Module
2. Transom
3. Mullion
4. Water barrier
5. L-profile support rail
6. Wall bracket
7. Insulation
8. Primary structure

Figure 5.25 Solaris 416. Axonometric drawing of the BIPV roof installation. Credits: adapted by Ana Marcos Castro from Huggenbergerfries Architekten ©.

Glasberga residential district

LOCATION	**Södertälje, Sweden (59° 11′ 44″ N)**
USE	**Residential**
PV PRODUCER	**Advanced Solar Power**
ARCHITECT	**The Paradoumo Group**
INSTALLER	**Soltech Energy Sweden**
YEAR OF CONSTRUCTION	**2020**

Project description: Located near the Glasberga Lake and only 40 km west of Stockholm, this solar residential district consists of 21 newly constructed chain houses and 6 villas. All houses are designed with sustainability and well-being in mind. The architecture follows traditional style using wood-framed construction while incorporating advanced technologies such as a BIPV roof and an EV charger. The chain houses come in three different designs and floor plans (varying from 112 m² to 130 m²) while the villas are aimed for bigger families, with each having a floor area of 225 m².

BIPV system key characteristics: The 28 BIPV roofing systems have orientations from southeast to southwest and cover nearly one-quarter of the buildings' annual electricity demand, including part of the EV charging requirements. Each chain house has a 2.6 kW BIPV roof using the 85 W framed Soltech RooFmodule while each villa integrates a 3 kW BIPV roof using the 40 W frameless Soltech ShingEl product. The rectangular BIPV modules are seamlessly integrated with metallic roof tiles, enabling a variation of roof designs: from open-gable roofs to M-shaped and valley-and-gable ones.

900 kWh/m²
Annual horizontal solar irradiation

Humid continental – Dfb
Köppen climate classification

Roof
BIPV application

New
Construction

CdTe solar shingles
PV technology

75.8 kW
BIPV array nominal power

73.5 kWh/m²
Annual electricity generation per BIPV module area

104.4 MWh
Annual electricity generation

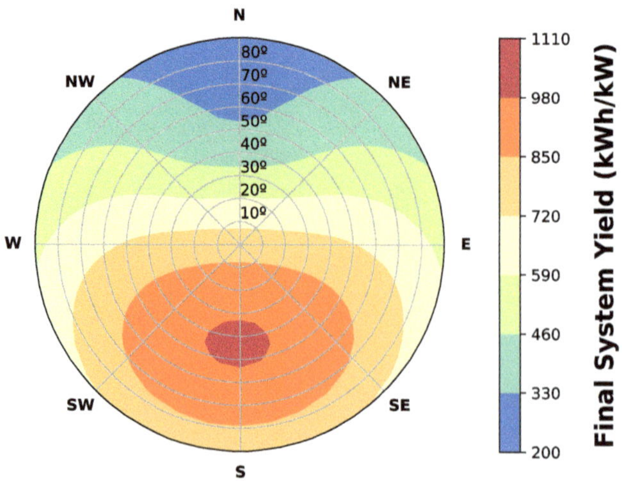

Figure 5.28 PV potential heatmap for Södertälje, Sweden. Credits: Ana Marcos Castro, CIEMAT.

Figure 5.26 Eagle-eye view of the Glasberga residential district. Credits: Soltech Energy Sweden.

Figure 5.27 Close-up view of the Glasberga BIPV roofs. Credits: Soltech Energy Sweden.

1. BIPV roofing

1

Not to scale

Figure 5.29 Glasberga. Plan view drawing. Credits: adapted by Ana Marcos Castro from Soltech Energy Sweden©. Credits: Soltech Energy Sweden.

Andreas Bjørns St.1

LOCATION	**Copenhagen, Denmark (55° 40′ 31″ N)**
USE	**Residential, Multi-Family**
CLIENT	**Owners Association**
PV PRODUCER	**Luxor (mounting system by Renusol)**
ARCHITECT	**Krydsrum Arkitekter**
ENGINEER	**Ekolab**
INSTALLER	**GAIA**
YEAR OF RENOVATION	**2013**

Project description: The building is a heritage multi-unit residential building located in the Christianshavn neighbourhood of Copenhagen, Denmark. The property, which covers a gross floor area of 1977 m², underwent a façade and roof renovation in 2013, resulting in a BIPV roof replacing the existing and outdated metal corrugated roof. Krydsrum Arkitekter with Ekolab were responsible for the design, and Enemærke and Petersen carried out the construction.

BIPV system key characteristics: The 240 m² roof was retrofitted with 108 opaque m-Si photovoltaic modules (1645 × 900 mm each) using tempered glass as the frontsheet and polyether on the back. The framed modules cover nearly 60% of the roof surface and are blended with conventional metal roof products of similar colour to create a uniform look from the street level. The BIPV area is 140 m² with a total 20,3 kW peak power, generating nearly 20% of the building's annual electricity consumption. The building includes additional energy conservation measures such as upgraded roof insulation and new high-performance windows.

1 MWh/m²
Annual horizontal solar irradiation

Marine West Coast – Csb
Köppen climate classification

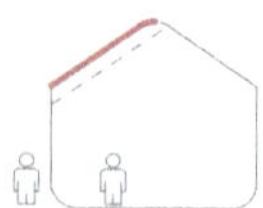

Roof
BIPV application

Retrofit
Construction

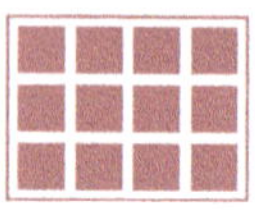

Opaque m-Si
PV technology

20.3 kW
BIPV array nominal power

121.4 kWh/m²
Annual electricity generation per BIPV module area

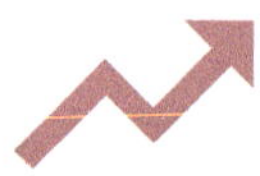

17 MWh
Annual electricity generation

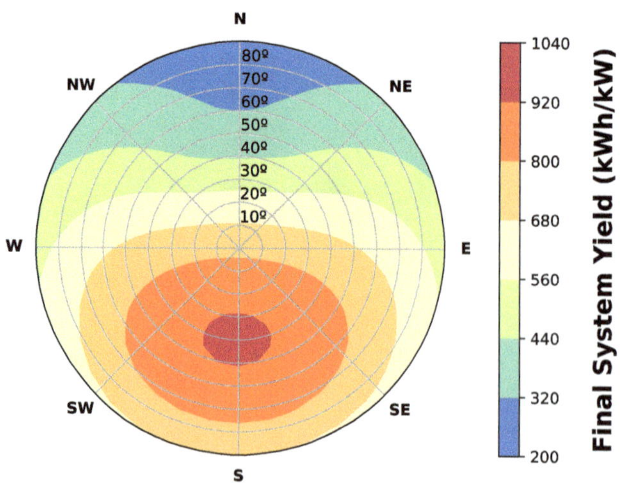

Figure 5.34 PV potential heatmap for Copenhagen, Denmark. Credits: Ana Marcos Castro, CIEMAT.

Figure 5.30 Andreas Bjørns St.1 renovated multi-unit residential building. Credits: Dorte Krogh.

Figure 5.31 Andreas Bjørns St.1 before renovation. Credits: Solar City Denmark.

Figure 5.32 Close-up of the BIPV roof at Andreas Bjørns St.1. Credits: Dorte Krogh.

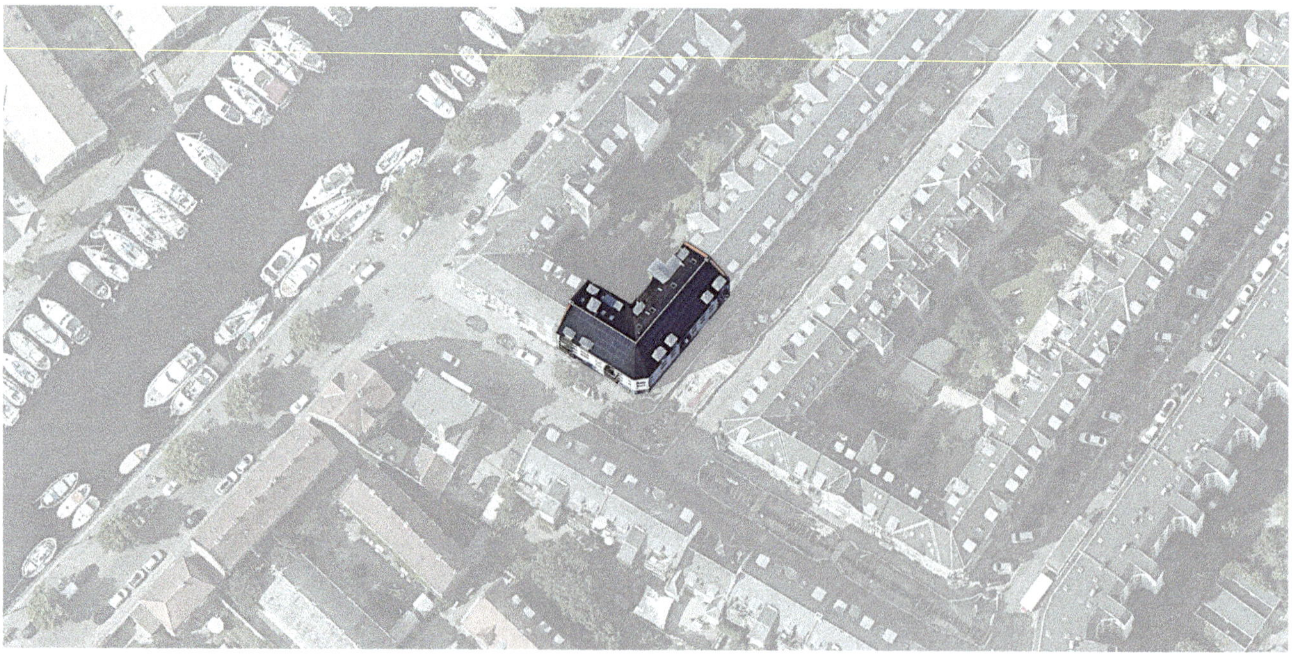

Figure 5.33 Site plan of Andreas Bjørns St.1. Source: Google Maps.

1. Framed PV module
2. Ventilated air cavity
3. Batten
4. Counter batten
5. EPDM membrane (waterproofing)
6. Roof sheathing
7. Rafter
8. Insulation

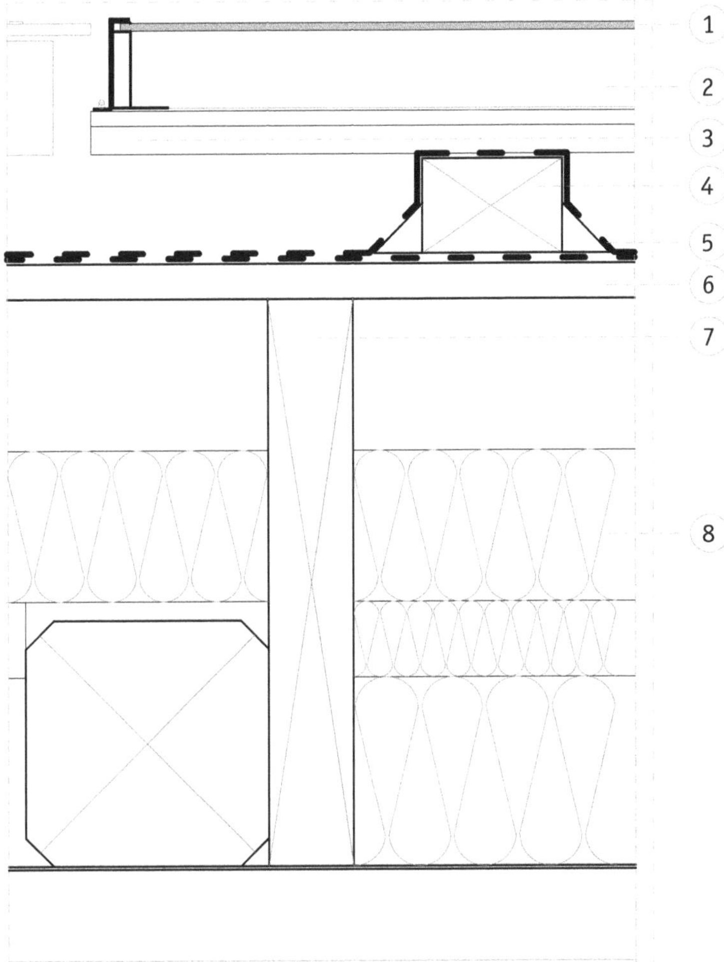

Section A-A'
Scale 1:4

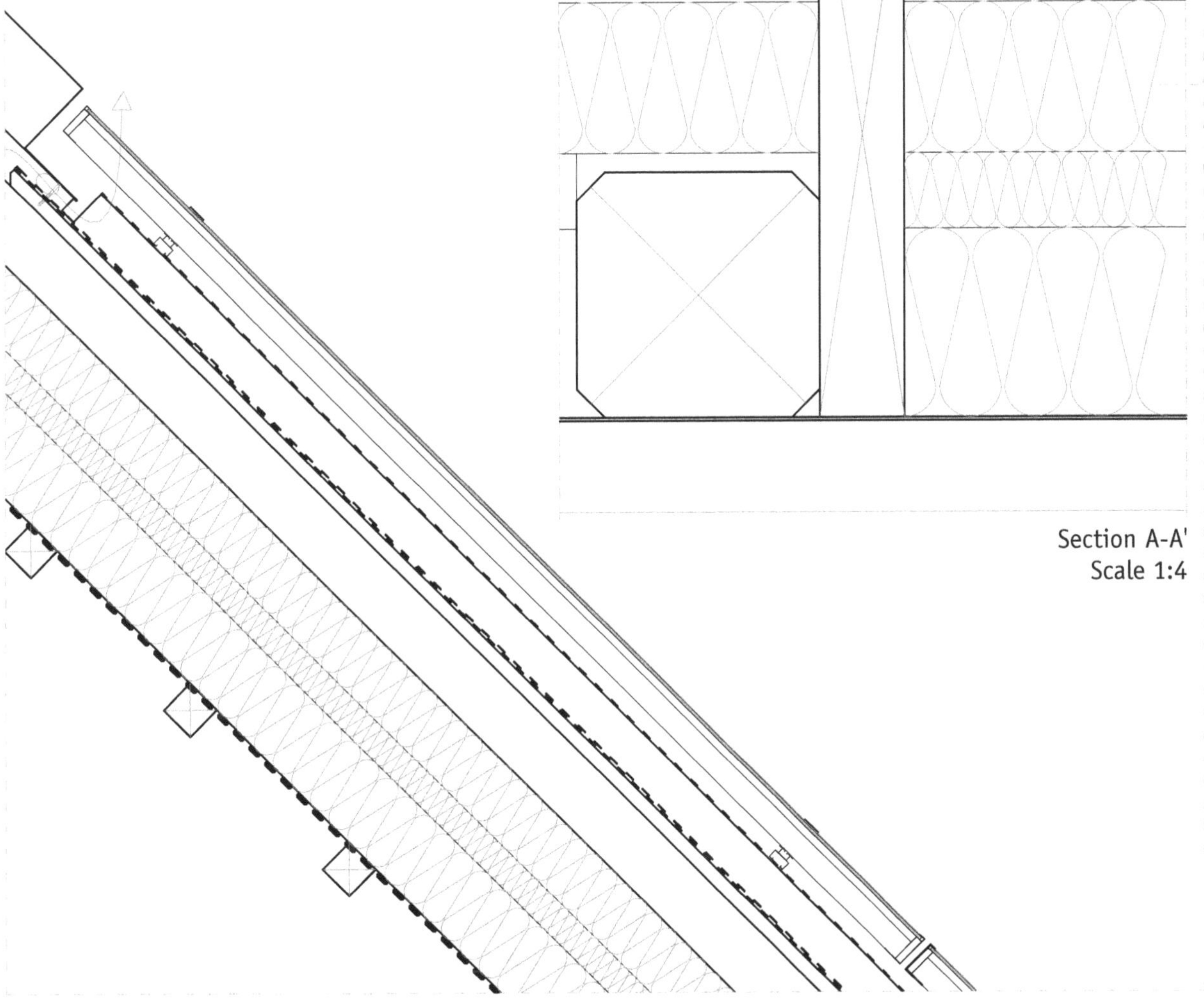

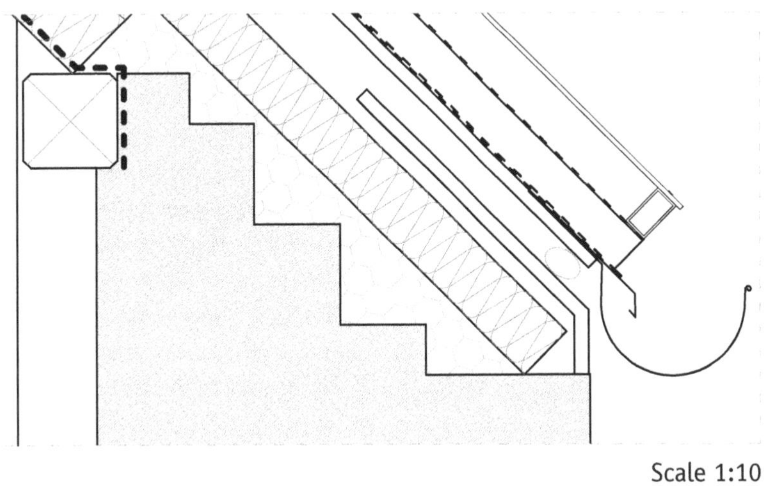

Scale 1:10

Figure 5.35 Andreas Bjørns St.1. Cross-section drawing of the BIPV roof installation. Credits: adapted by Ana Marcos Castro from Krydsrum Arkitekter ©.

Urban park in Certosa Island

LOCATION	**Venice, Italy (45° 25 ′58″ N)**
USE	**Multi-purpose**
PV PRODUCER	**EnergyGlass TG-EGM²4ST and TG-EGM45ST**
ENGINEER	**Solmonte S.r.L., Ing. Sofia Tiozzo Pezzoli**
INSTALLER	**Solmonte S.r.l.**
YEAR OF RENOVATION	**2020**

Project description: The Urban Park is located on Certosa Island, in the northern Venetian lagoon. This 22-hectare island is a UNESCO heritage site. Since the 1990s, it has been the focus of a development initiative that involved turning vacant land into an urban park. Public-private cooperation between the municipality and local developers has resulted in the restoration of several buildings. Three non-protected buildings have been retrofitted with BIPV roof tiles, respecting the historical value of the building and the heritage site in which it is inserted.

BIPV system key characteristics: The BIPV roof panels cover a total surface area of 1,100 m². With a tilt angle of 18° to 25°, the glass-glass-coloured modules are attached to the roof deck using a system of brackets, channels, and drains to ensure water tightness. The use of glass/glass frameless terracotta-coloured modules enabled the harmonious integration of a disruptive building envelope technology such as BIPV with the traditional architecture of Northern Italy.

1.1 MWh/m²
Annual horizontal solar irradiation

Humid subtropical – Cfa
Köppen climate classification

Roof
BIPV application

Retrofit (UNESCO heritage site)
Type of construction

Opaque coloured m-Si
PV technology

183 kW
BIPV array nominal power

192 kWh/m²
Annual electricity generation per BIPV module area

211 MWh
Annual electricity generation

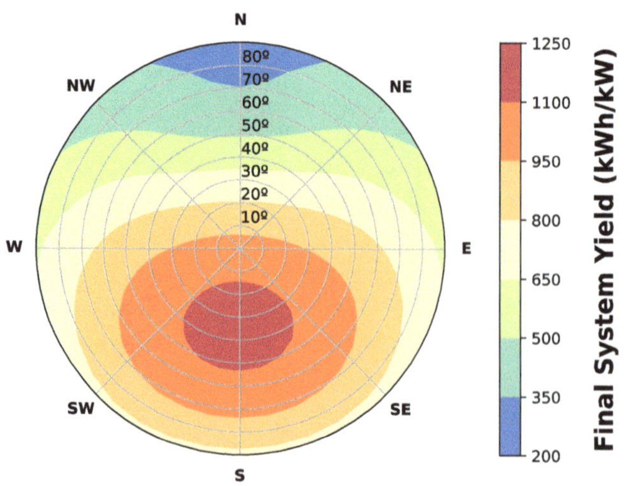

Figure 5.39 PV potential heatmap for Venice, Italy. Credits: Ana Marcos Castro, CIEMAT.

Figure 5.36 BIPV roof in a heritage building at the Urban Park in Certosa Island. Credits: GruppoSTG.

Figure 5.37 Detail of the terracotta BIPV roof. Credits: GruppoSTG.

Figure 5.38 Site plan of the Urban Park in Certosa Island. Source: Google Maps.

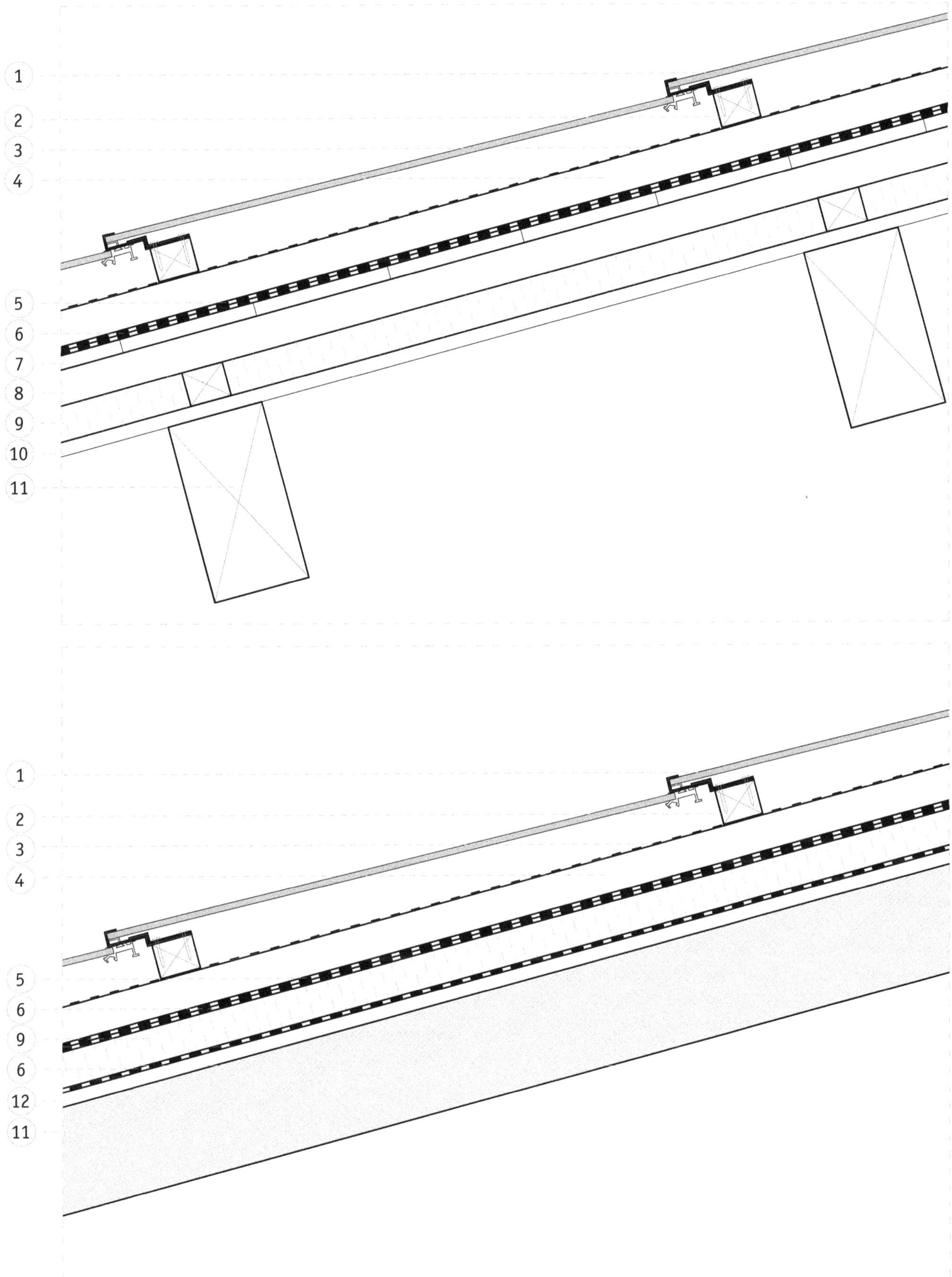

Scale 1:8

1. BIPV module
2. Batten
3. Vapor barrier
4. Counter batten
5. New bituminous water membrane
6. Existing water membrane
7. OSB sheathing
8. Rafter
9. Rigid insulation
10. Existing roof sheathing
11. Primary structure
12. Mortar

Figure 5.40 Urban Park in Certosa Island. Cross-section drawing of the BIPV roof installation. Credits: adapted by Ana Marcos Castro from Solmonte ©. Credits: Solmonte.

5.2.2 Skylight

General: BIPV skylights frequently act as prominent architectural features, establishing visually stunning and welcoming entrances, courtyards, and atria flooded with daylight. BIPV skylight installations typically use curtain wall solutions with special considerations to electrical, structural, and architectural requirements. This section discusses design aspects specific to BIPV skylight installations. Complementary information can be found under BIPV curtain wall systems. Figure 5.41 shows a 3D image of a cropped construction section of a skylight. Labels are explained as follows.

Structural Design: BIPV skylights are exposed to various loads, including dead loads, environmental loads such as wind, snow, or ice, and maintenance loads. The framing, the connections ① to the main structure, and the BIPV insulated glass units (IGU) must be carefully engineered to safely support these loads while conforming to local building codes and standards. The use of heat-strengthened or fully tempered laminated innermost glass is recommended to enhance safety from glass breakage. Framing elements ② should be designed to account for deflection and thermal expansion. When curtain wall solutions are implemented, the mullions perform the role of rafters while the transoms act as purlins. The use of splice joints can accommodate for the frame movement due to wind and thermal variations. For long-span skylights, the use of a reinforcing superstructure such as steel ribs and cable nets can increase stiffness and add structural integrity and durability. Note that the maximum deflection of the BIPV frame is determined based on the deflection limits imposed by the IGUs ③. For inspiration, see Porta Susa station in Turin and Azurmendi restaurant near Bilbao, respectively.

Hygrothermal Performance: Similarly to BIPV curtain walls, special attention should be given to moisture (water and condensation) control. Depending on the skylight design, the pressure-equalised cavity ④ can be slightly wider than in traditional curtain wall systems to house the DC wiring ⑥. Any moisture that enters the cavity is drained out by gravity.

Fire Safety: The local or regional fire codes may require an automatic sprinkler protection system and an automatic or manual DC disconnect, depending on the size, location, and fire-resistance rating of the BIPV skylight system and products used.

Electrical Considerations: When using an edge junction box with BIPV IGUs, DC wiring can be routed along the mullions and transoms. Conversely, when a conventional rear junction box is used, wiring can run indoors, concealed within trays, for cable management and aesthetic reasons.

Architectural Considerations: BIPV skylights can be used either in new or existing buildings. Low-slope skylights (15° or less) can be effective daylighting solutions for locations with latitude between 0° and 45°. On the contrary, low solar angles in locations with latitude 45° or higher limit the daylight and PV generation effectiveness of low-slope skylights [7]. For these locations, higher skylight slopes are recommended (i.e., latitude ± 15°). The choice of PV technology, colours, and transparency levels creates an opportunity for the building designers to produce a stimulating luminous environment, enhance the user experience, and transform the BIPV skylight into a focal architectural feature. The use of coloured thin-film semi-transparent BIPV products can flood an atrium with daylight and colour. The use of Si-based products incorporating spaced opaque solar cells can produce ever-changing patterns of light and shadows while the use of translucent products can create a uniform luminous environment.

Sustainability: A BIPV skylight should be designed to strike a balance between solar electricity generation and control of solar heat gains and glare. Its direct and indirect impact on building energy consumption (i.e., heating, cooling, lighting and electricity generation) and indoor environmental quality (i.e., visual, thermal and acoustic comfort, and indoor air quality, when natural ventilation is incorporated) should be assessed and optimised at the design development stage, using building performance simulations.

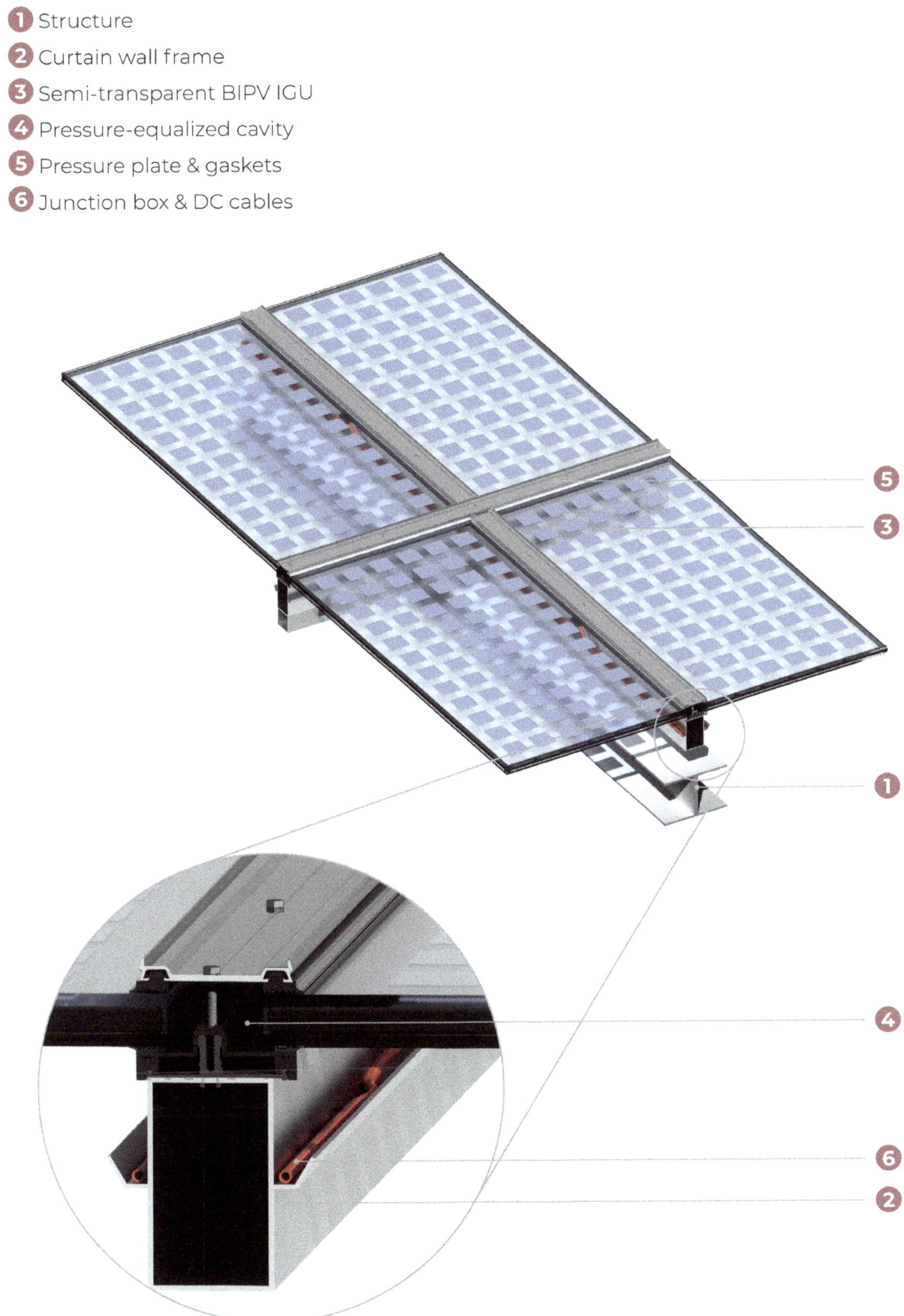

Figure 5.41 Axonometric representation of a typical BIPV skylight. Credits: Marley Dowling.

San Antón Market

LOCATION	Madrid, Spain (40° 25′ 0″ N)
USE	Farmers' market
PV PRODUCER	L Vision (20%) by Onyx Solar
ARCHITECT	estudio ATARIA
ENGINEER	estudio ATARIA
INSTALLER	Onyx Solar
YEAR OF RENOVATION	2010

Project description: Located in the heart of Chueca district, San Antón Market is one of the most vibrant markets in central Madrid. Its history dates to the 19th century. The market was renovated in 2011 (and more recently in 2021). Spreading over three floors and with a gross floor area of 7,345 m², the new market provides a wide range of fresh produce, food vendors, and restaurants, playing an integral part in the daily life of the Madrilenians, in one of the most densely populated districts of the Spanish capital. A centrally located photovoltaic skylight brings daylight to all floors while supplying the market with renewable electricity.

BIPV system key characteristics: The BIPV skylight system has a total surface of 168 m². It comprises 54 BIPV insulated glazing units in a saw-tooth arrangement. Each BIPV unit has an *SHGC* of 12%, visible transmittance of 20%, and a nominal power output of 94 W. The junction boxes and wiring are neatly hidden within the framing system of the skylight. The return on investment for the BIPV skylight was less than 2 years.

1.7 MWh/m²
Annual horizontal solar irradiation

Hot-summer Mediterranean – Csa
Köppen climate classification

Skylight
BIPV application

Retrofit
Construction

Semi-transparent amorphous Si
PV technology

6.5 kW
BIPV array nominal power

46 kWh/m²
Annual electricity generation per BIPV module area

7.7 MWh
Annual electricity generation

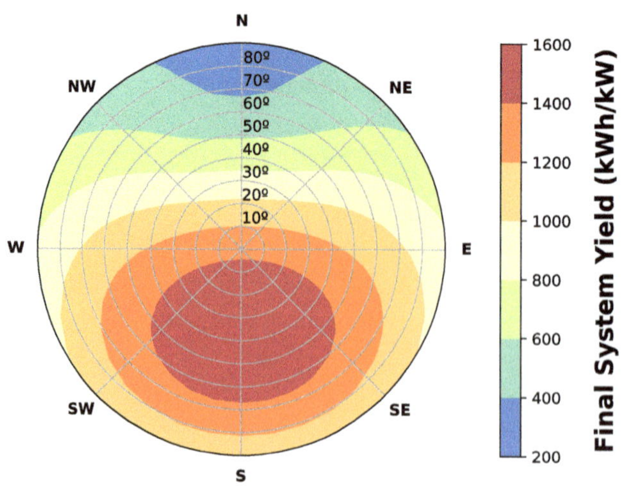

Figure 5.45 PV potential heatmap for Madrid, Spain. Credits: Ana Marcos Castro, CIEMAT.

Figure 5.42 Pedestrian view of San Antón Market. Credits: Nuria Martin Chivelet, CIEMAT.

Figure 5.43 Interior view of the BIPV skylight at San Antón Market. Credits: Onyx Solar.

Figure 5.44 Exterior view of the BIPV skylight at San Antón Market. Credits: Onyx Solar.

Figure 5.46 Site plan of San Antón Market. Credits: Google Maps.

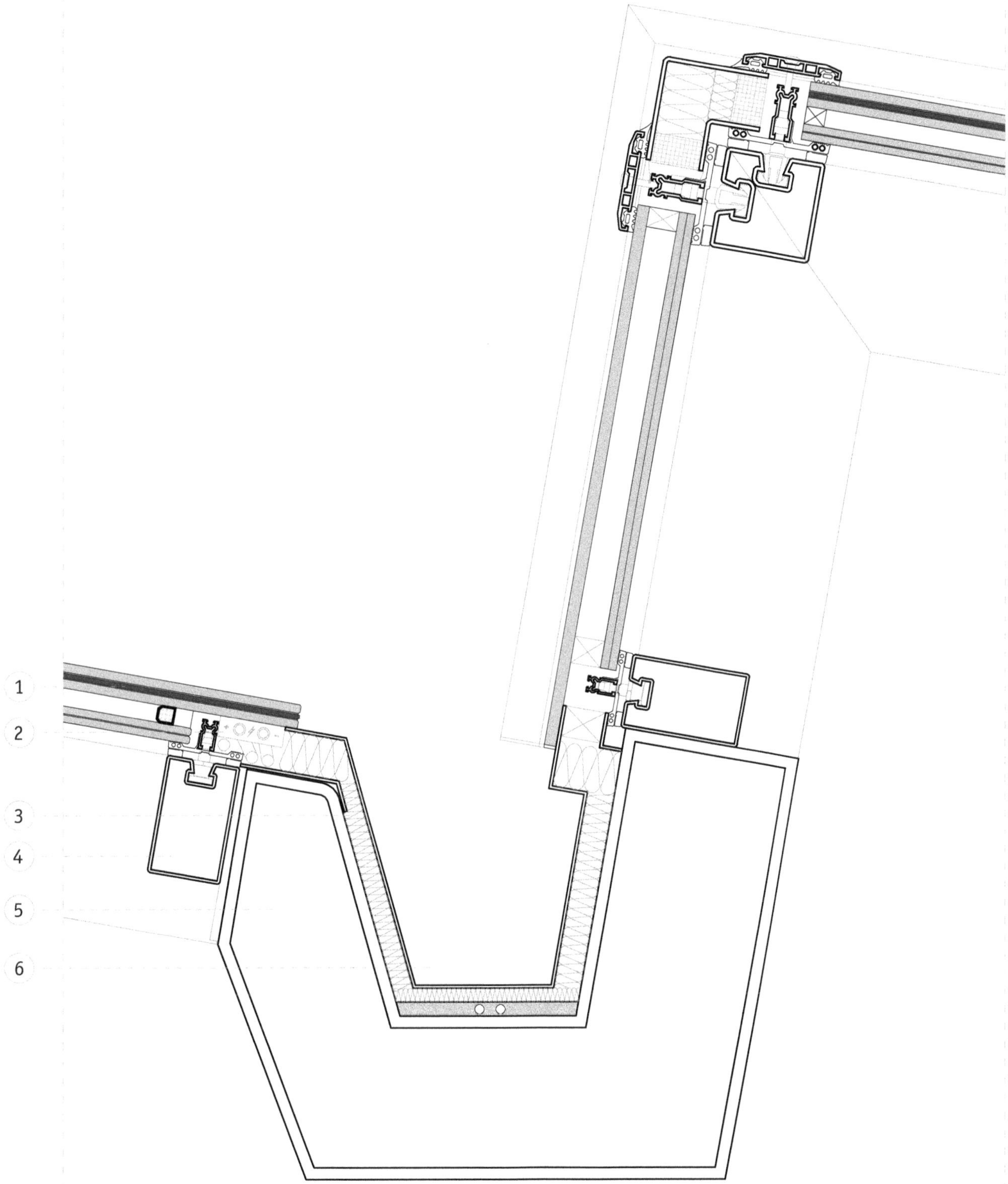

1. BIPV module
2. Junction box and wiring
3. Insulation
4. Steel transom
5. Steel beam
6. Integrated gutter

Figure 5.47 San Anton Market. Cross-section drawing of the BIPV roof installation. Credits: adapted by Ana Marcos Castro from Onyx Solar ©.

Scale 1:4

Azurmendi Restaurant

LOCATION	Bilbao, Spain (43° 15′ 38″ N)
USE	Restaurant
PV PRODUCER	L Vision (20%) by Onyx Solar
ARCHITECT	Naia Eguino
ENGINEER	Proiek Enginyeria SL
INSTALLER	Onyx Solar
YEAR OF RENOVATION	2015

Project description: The three Michelin stars restaurant headed by Eneko Atxa is located at the Biscayan town of Larrabetzu, near the city of Bilbao. The building is half-buried in a hillside with unobstructed views to the valley. In 2014, the building obtained the LEED Gold certification and was declared as the most sustainable restaurant in the world.

BIPV system key characteristics: The BIPV installation is composed of a curtain wall and a skylight system with over 270 m² of photovoltaic glass. The optical properties of the BIPV IGUs were carefully selected to balance daylight utilisation and solar heat gains (*SHGC* of 9% and 12% for skylight and curtain wall IGUs, respectively). The final design results in a reduction of 55% on building energy consumption, which corresponds to a reduction of 11 metric tons of CO_2 emissions per year.

1.2 MWh/m²
Annual horizontal solar irradiation

Humid temperate – Cfb
Köppen climate classification

Skylight and curtain wall
BIPV application

New
Type of construction

Semi-transparent amorphous Si
PV technology

21 kW
BIPV array nominal power

66 kWh/m²
Annual electricity generation per BIPV module area

16.4 MWh
Annual electricity generation

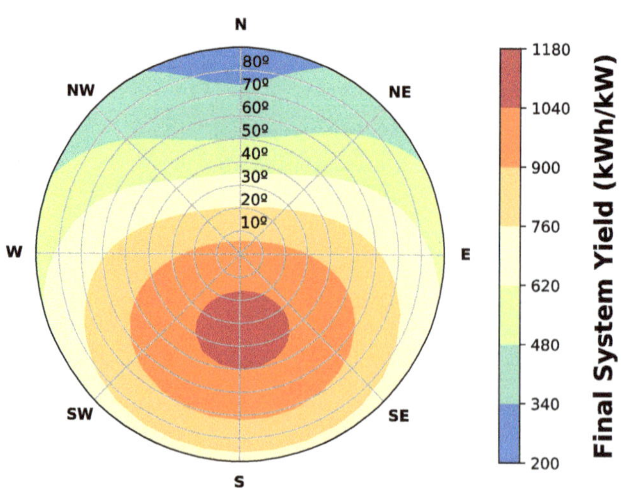

Figure 5.52 PV potential heatmap for Bilbao, Spain. Credits: Ana Marcos Castro, CIEMAT.

Figure 5.48 General view of Azurmendi Restaurant showing the BIPV façade and skylight. Credits: Onyx Solar.

Figure 5.49 Interior view of BIPV curtain wall and skylight at the Azurmendi Restaurant. Credits: Onyx Solar.

Figure 5.50 Semitransparent curtain wall at the Azurmendi Restaurant. Credits: Onyx Solar.

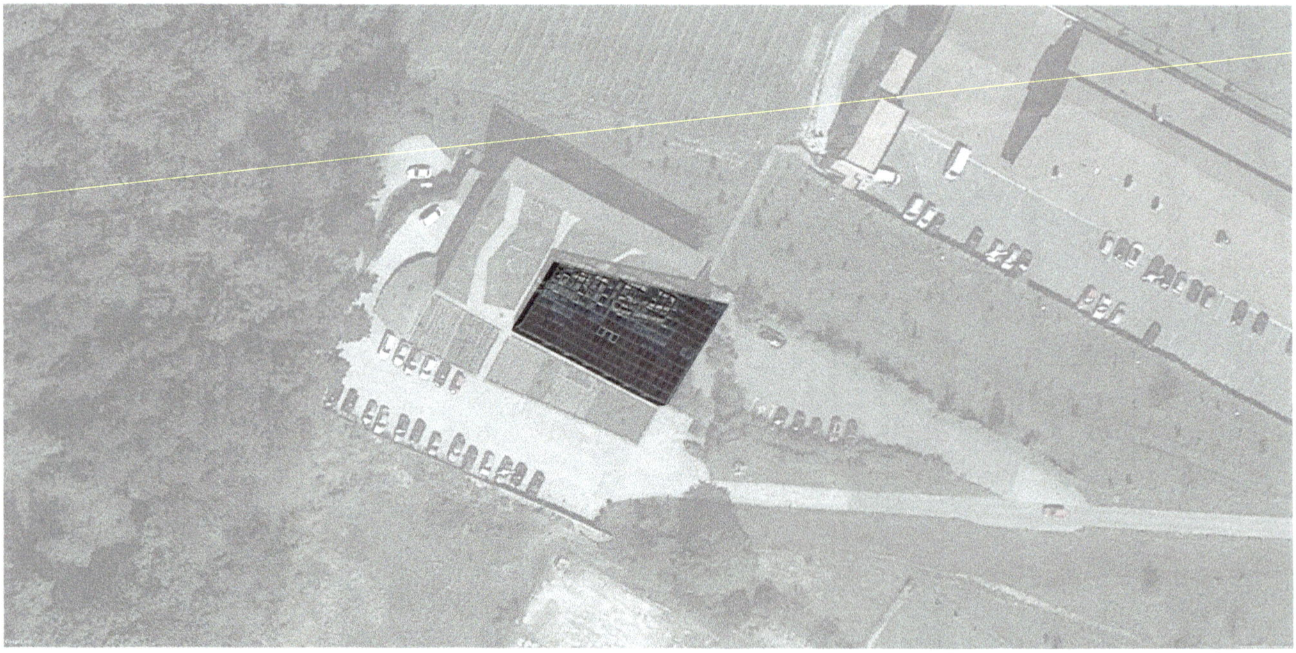

Figure 5.51 Site plan of Azurmendi Restaurant. Source: Google Maps.

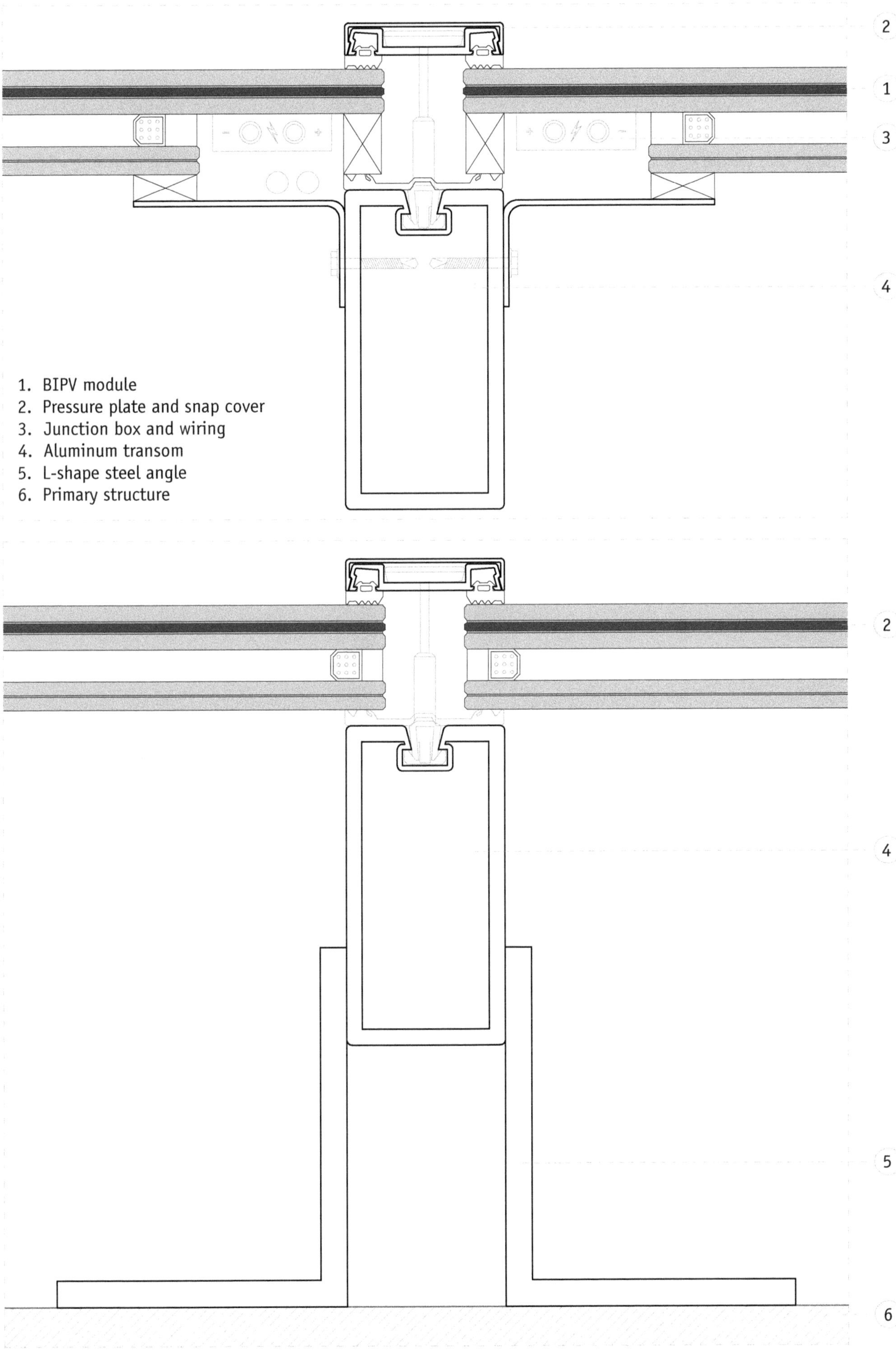

Figure 5.53 Azurmendi Restaurant. Cross-section drawing of the BIPV roof installation. Credits: adapted by Ana Marcos Castro from Onyx Solar ©.

Edmonton Convention Centre

LOCATION	**Edmonton, Canada (53° 32′ 31″ N)**
USE	**Institutional**
PV PRODUCER	**Onyx Solar**
ARCHITECT	**DIALOG**
ENGINEER	**Howell-Mayhew Engineering**
INSTALLER	**Kuby Energy**
YEAR OF RENOVATION	**2020**

Project description: Located in the city of Edmonton, Alberta, the newly retrofitted Edmonton Convention Centre is home to Canada's largest BIPV system. Inaugurated in 2020, the renovated multi-level atrium of the convention centre had all 696 of its sloped glass panels replaced with photovoltaic semi-transparent modules. The installation has a total area of 1,566 m².

BIPV system key characteristics: The BIPV modules are installed at a variety of orientations and tilt angles due to the design of the atrium. More than ten different types of PV glass were used in this project. This project was an important contribution to the City of Edmonton's strategic goals toward sustainability, energy efficiency, and art promotion.

BIPV modules were chosen to improve the energy efficiency of the building and reduce its carbon footprint, as well as enhance the aesthetics of the building. This was achieved through the design of the 'Oculus', a section of the installation where BIPV modules were used to imprint an excerpt from the poem 'Gifts of a River' by former Edmonton poet Laureate E.D. Blodgett in Morse code. The 'Oculus' also increases the overall light transmittance in the atrium and meets the Percent Art Program of the City of Edmonton.

1.3 MWh/m²
Annual horizontal solar irradiation

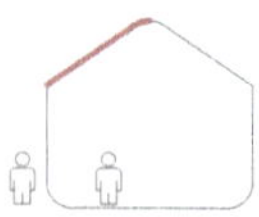

Skylight
BIPV application

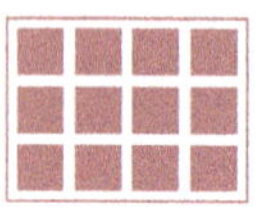

Semi-transparent m-Si
PV technology

128 kWh/m²
Annual electricity generation per BIPV module area

Humid continental- Dfb
Köppen climate classification

Retrofit
Type of construction

169 kW
BIPV array nominal power

200 MWh
Annual electricity generation

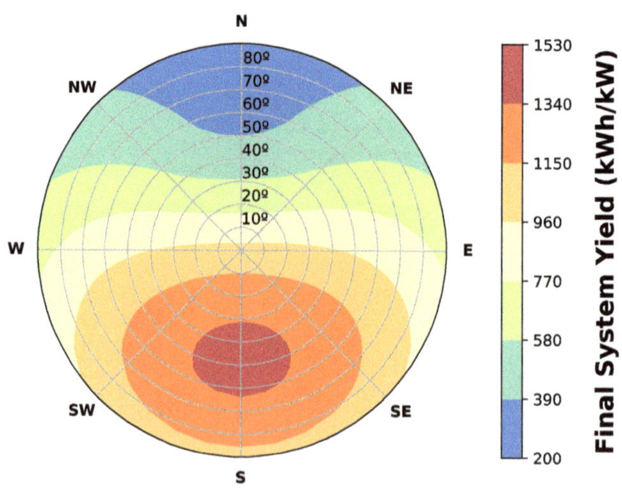

Figure 5.58 PV potential heatmap for Edmonton, Canada. Credits: Ana Marcos Castro, CIEMAT.

Figure 5.54 General view of Edmonton Convention Centre. Credits: Onyx Solar.

Figure 5.55 Interior view of the BIPV skylight at Edmonton Convention Centre. Credits: Onyx Solar.

Figure 5.56 Skylight oculus at Edmonton Convention Centre. Credits: Onyx Solar.

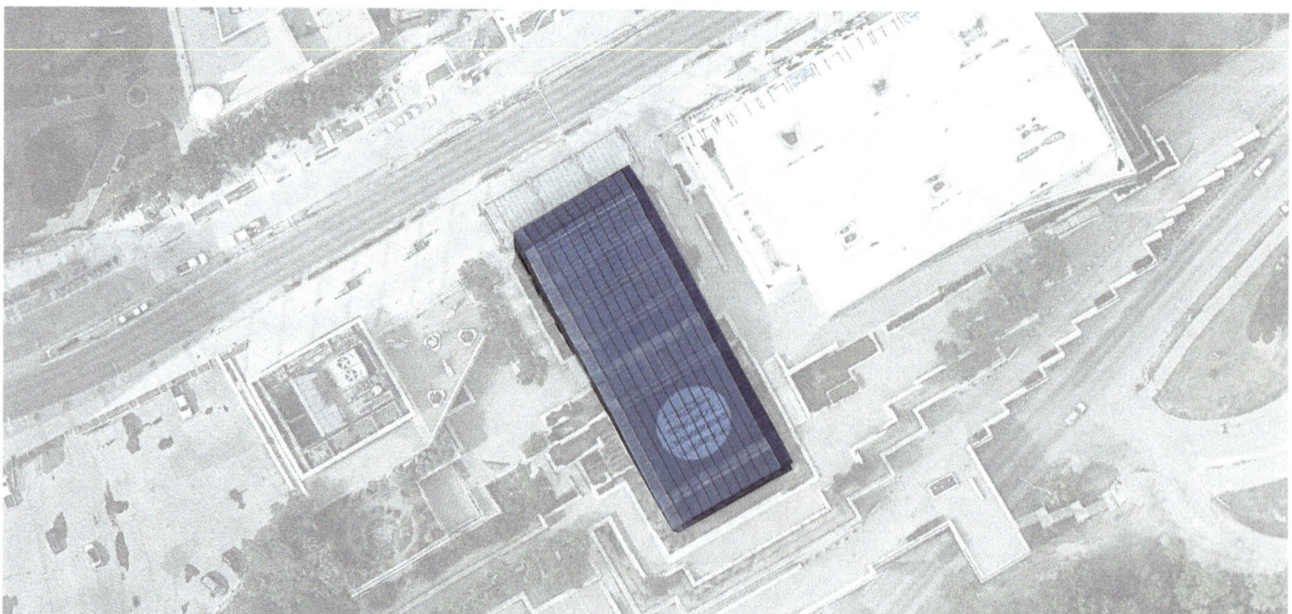

Figure 5.57 Site plan of Edmonton Convention Centre. Source: Google Maps.

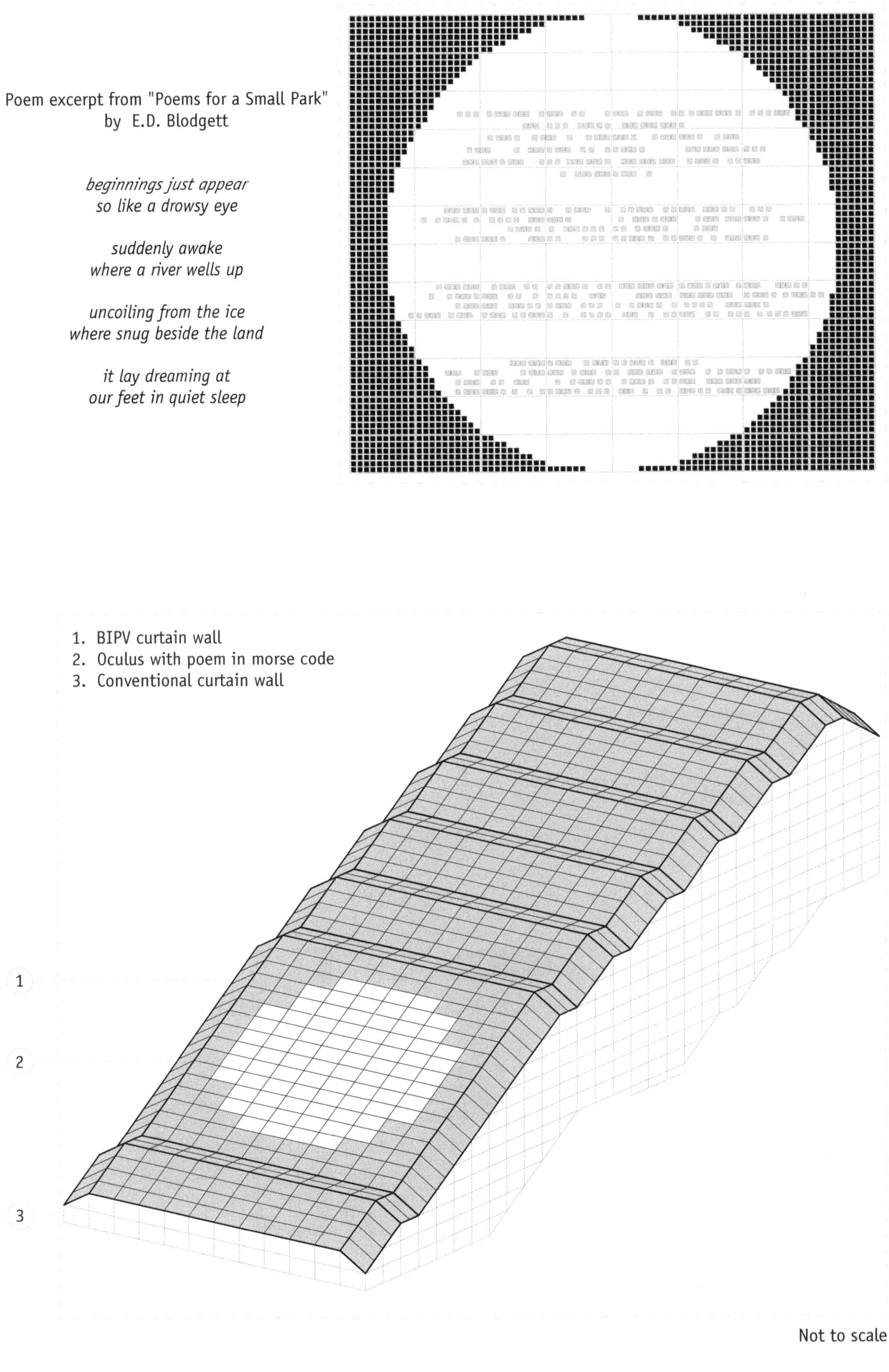

Figure 5.59 Edmonton Convention Centre. Axonometric drawing of the renovated BIPV roof installation and oculus design. Credits: adapted by Ana Marcos Castro from Onyx Solar ©.

Porta Susa AV Station

LOCATION	Turin, Italy (45° 4′ 22″ N)
USE	Train station
PV PRODUCER	EnergyGlass, GruppoSTG
ARCHITECT	AREP, Silvio d'Ascia Architecture, Agostino Magnaghi
ENGINEER	AREP
INSTALLER	GruppoSTG
YEAR OF CONSTRUCTION	2013

Project description: Located in Turin city centre, Porta Susa is the first stop on the Paris-Rome line and is considered a main "port of entry" for trains coming from Northern Europe. The 15,000 m² glass tunnel on a steel superstructure extends 386 m × 30 m, at varying heights. The tunnel encloses but also connects the railway and commercial space with the surrounding city.

BIPV system key characteristics: The semi-transparent BIPV modules cover approximately 9,000 m² of the tunnel's glass surface. The BIPV system supplies the station with solar electricity while allowing daylight to illuminate the train platforms and commercial space. The BIPV modules provide solar shading and an appealing visual indoor environment by alternating opaque solar cells and transparent glass. Considering the variable orientation of the BIPV modules on the curved surface of the glass tunnel, the BIPV modules connected in series under the same string are wired to a maximum of five rows in order to minimise energy generation losses due to variable module orientation.

1.1 MWh/m²
Annual horizontal solar irradiation

Humid subtropical – Cfa
Köppen climate classification

Skylight
BIPV application

New
Type of construction

Semi-transparent m-Si
PV technology

765 kW
BIPV array nominal power

75 kWh/m²
Annual electricity generation per BIPV module area

680 MWh
Annual electricity generation

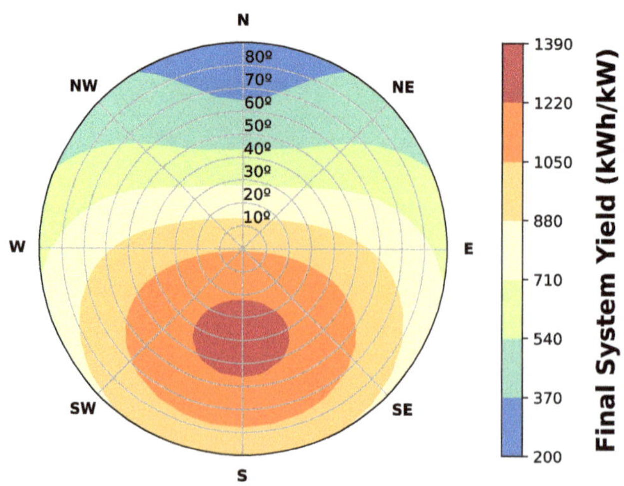

Figure 5.64 PV potential heatmap for Turin, Italy. Credits: Ana Marcos Castro, CIEMAT

Figure 5.60 Eagle-eye view of Porta Susa AV Station. Credits: Michele d'Ottavio.

Figure 5.61 Pedestrian interior view of the BIPV skylight at Porta Susa AV Station. Credits: Julien Lanoo.

Figure 5.62 Detail of the BIPV skylight and dummy modules. Credits: Francesco Frontini, SUPSI.

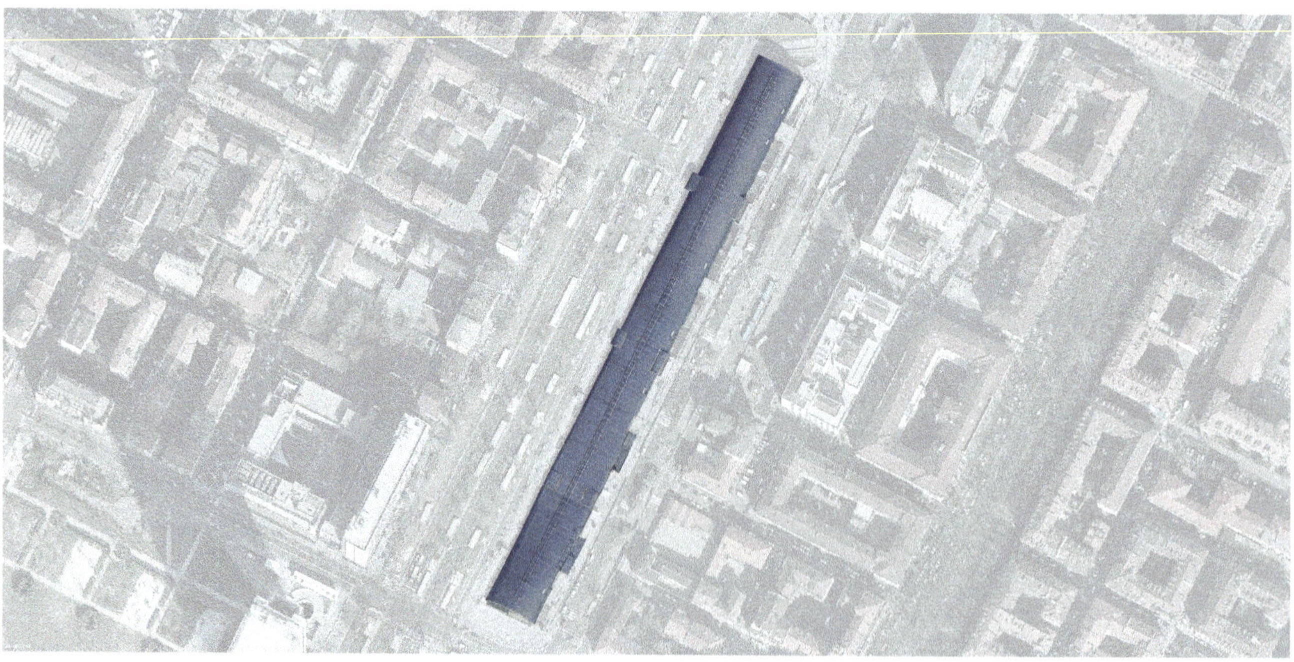

Figure 5.63 Site plan of Porta Susa AV Station. Source: Google Maps.

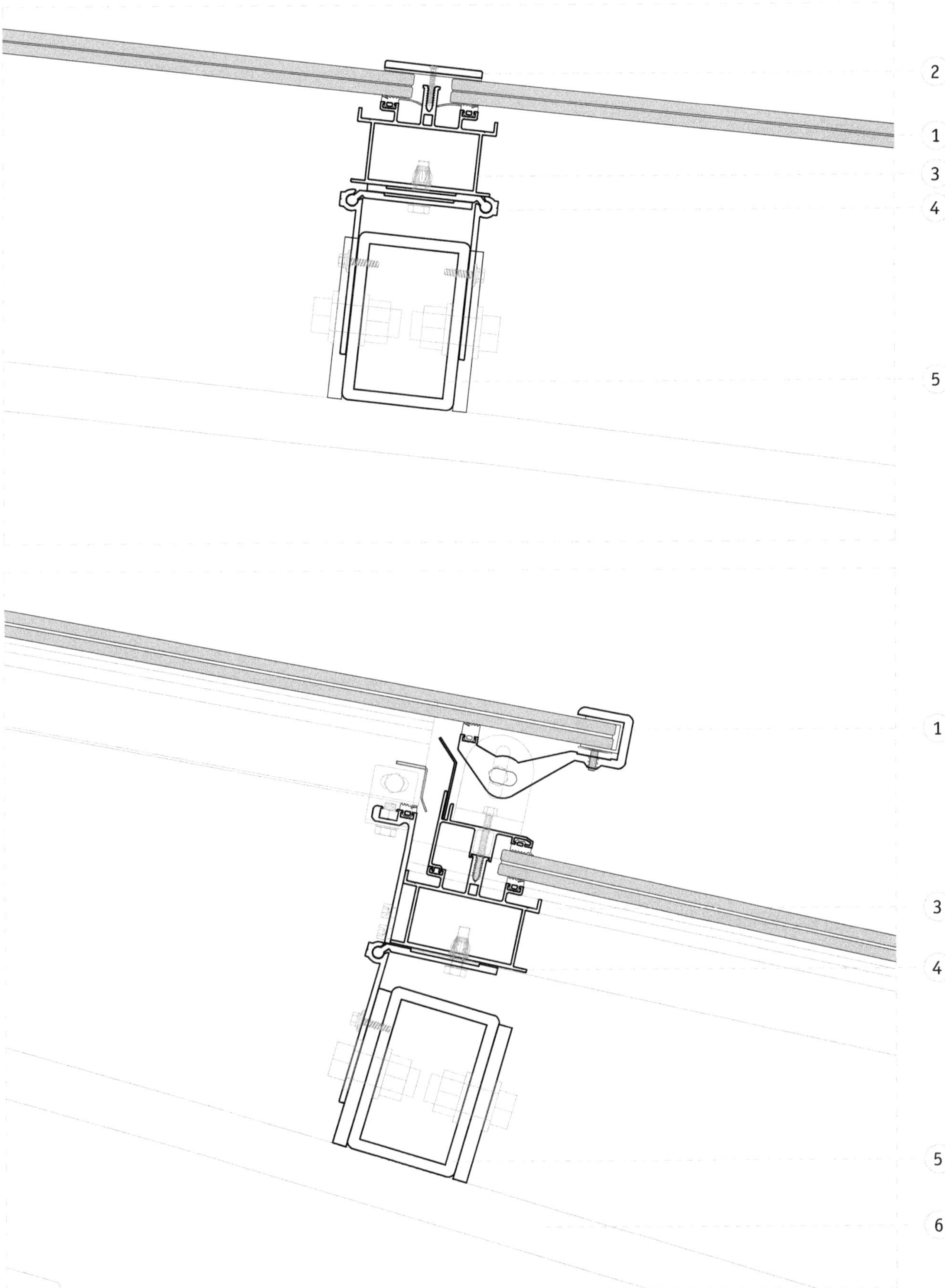

1. BIPV module
2. Pressure plate
3. Aluminum profile
4. Hinge plate
5. Transom
6. Steel structure

Figure 5.65 Porta Susa AV Station. Cross-section drawings of the renovated BIPV roof installation Source: adapted by Ana Marcos Castro from Gruppo STG ©.

5.3 BIPV façade systems

5.3.1 Rainscreen

General: A BIPV rainscreen can be a cost-effective envelope solution for both new-build and retrofit projects. Its design considerations align with those of conventional rainscreen systems, supplemented by electrical and fire safety requirements [8].

Structural Design: The BIPV rainscreen must be engineered to endure and transfer both dynamic and static loads to the load-bearing backing wall ①, in compliance with the local building codes. These primarily include the dead load resulting from the weight of the BIPV modules and subframe members, as well as dynamic wind loads. The connections between the brackets ③ and the vertical rails ⑤ are typically made with self-drilling screws or bolts and nuts, using fixed or sliding points. The sliding points are used to allow thermal expansion of the rails. The BIPV modules ⑫ can be mounted on the horizontal rails ⑥ using visible compression fittings ⑩ or concealed anchors adhered to the rear side of the BIPV modules. BIPV rainscreens tend to be drained and ventilated with open joints between modules to increase rear air flow while also allowing thermal expansion of the modules. When using a steel structure (galvanised or stainless) with framed BIPV modules, it is important to avoid electrical contact between metals and prevent galvanic corrosion of the anodised aluminium frame. This is not of concern with frameless BIPV modules.

Hygrothermal Performance: The BIPV cladding plays the role of the primary water barrier, deflecting most of the direct rain. A rainscreen cavity ⑧ allows any penetrating moisture to drain or evaporate and vent to the outside. The rainscreen cavity should be on average 100 mm or more to ensure adequate natural air flow for the BIPV modules and avoid overheating. Parapet and drain flashing should permit air flow while the installation of a permeable insect screen at the top and bottom course of rainscreen is recommended. An additional water/air barrier ② is applied (e.g., adhered membrane or spray applied) on the exterior face of the backing wall to eliminate any further water ingress and ensure an airtight construction. If required, electrical conduit penetrations should be used to bring the DC wiring from the rainscreen to the electrical room, using a flexible silicone grommet that offers an airtight and watertight seal, while also allowing the conduit to move without compromising the seal. The choice of thermal insulation ④ material and thickness depends on local building codes, fire safety regulations, and sustainability needs. Thermally broken brackets ③ should be used to reduce thermal bridging.

Fire Safety: The use of non-combustible insulation (e.g., mineral wool) is advised to reduce or prevent fire propagation through the rainscreen cavity. The use of fire barriers with an intumescent strip installed across the cavity can be an additional fire and smoke control practice.

Electrical Considerations: Effective cable management and labeling ensure BIPV rainscreen longevity. The use of cable trays ⑦, clips, and ties can help secure the wiring, prevent mechanical stress, and avoid abrasion from sharp corners and movement due to wind or other vibrations that could damage wiring or connectors. The connectors should have a minimum rating of IP 65 while the DC cables should be installed to avoid induction loops that can generate significant magnetic fields. When possible, the modules should be connected in a leapfrog fashion (vs daisy chain wiring). Façade shading should be accounted for during the string design. When applicable, low-profile pressure caps ⑪ are recommended to avoid casting shadows to the BIPV modules under high solar incidence angles.

Architectural Considerations: The use of coloured BIPV enables architects and designers to seamlessly integrate solar panels into the building's design, as the BIPV rainscreen can be customised to match or complement the building's colour scheme and architectural style, creating visually appealing and innovative structures that stand out while still being sustainable. When the BIPV rainscreen is susceptible to impacts or damage, protection can be ensured by taking precautions such as installing the modules at least 0.8 m above the ground or above the wall skirting.

Sustainability: The BIPV rainscreen should be engineered to last for at least 35 years with future reuse in mind. Designing an adaptable sub-frame that accommodates different types of modules enables the building envelope to adapt to changing technologies or aesthetic preferences. The system design should also consider ease of maintenance. Components that require periodic maintenance or replacement should be easily accessible without disrupting building operations.

An axonometric of a typical BIPV rainscreen. The drawings illustrate all key components of a BIPV rainscreen system.

1. Load-bearing backing wall
2. Water/air barrier, adhered membrane or spray applied
3. Wall bracket, thermally broken
4. Thermal insulation, mineral wool
5. Vertical substructure rail
6. Horizontal substructure girt
7. Cable management tray
8. Drainage/air flow cavity, 80 mm or more
9. BIPV module, frameless
10. Pressure plate
11. Pressure cap
12. BIPV module junction box and DC cables

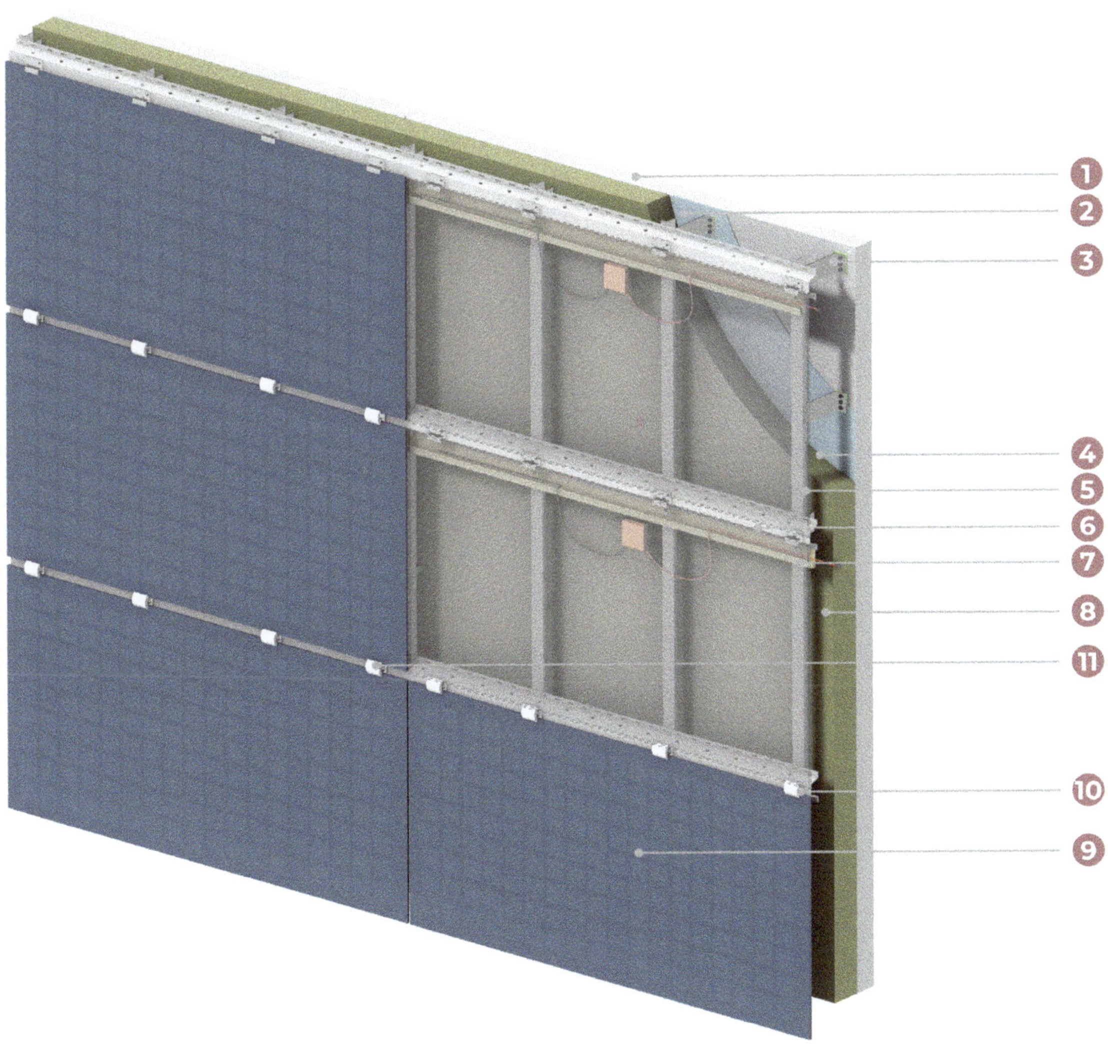

Figure 5.66 Axonometric representation of a typical BIPV rainscreen system. Credits: Sara Eskandar. Ana Marcos Castro, Marley Dowling, Amy Markwart.

Jeanne & Peter Lougheed Performing Arts Centre

LOCATION	Camrose, Canada (53° 0′ 46″ N)
USE	Concert Hall
PV PRODUCER	Coenergy
ARCHITECT	BR2 Architecture
ENGINEER	Howell-Mayhew Engineering & SolNorth
INSTALLER	Great Canadian Solar
YEAR OF CONSTRUCTION	2014

Project description: The Jeanne & Peter Lougheed Performing Arts Centre is a unique performing arts venue located on the University of Alberta's Augustana Campus in Camrose, Alberta. Inaugurated in 2014, this institutional building covers a total area of approximately 4,100 m² and has achieved a Green Globes rating of four globes. The BIPV installation is composed of 488 PV modules, with a total area of 882 m².

BIPV system key characteristics: The BIPV modules comprise the north, east, west, and south façades of the theatre's flytower. Modules were installed on the north façade, despite the low solar energy generation potential on that side, to provide the building with a uniform appearance.

The decision to use BIPV modules was driven by the GHG emission reduction plan by the University of Alberta, and it was also seen as an opportunity to educate students and create awareness on PV technologies. Furthermore, a BIPV system was chosen over a conventional building-applied system to reduce costs, as the slightly higher capital costs that went into installing the flytower cladding are offset by electricity bill savings. The BIPV system supplies approximately 20% of the building's energy requirements. At the time of installation, the estimated payback time was 10–15 years, while the expected lifetime of the system is expected to be over 40 years.

1.2 MWh/m²
Annual horizontal solar irradiation

Humid continental- Dfb
Köppen climate classification

Rainscreen
BIPV application

New
Type of construction

Opaque m-Si
PV technology

122 kW
BIPV array nominal power

185 kWh/m²
Annual electricity generation per BIPV module area

75 MWh
Annual electricity generation

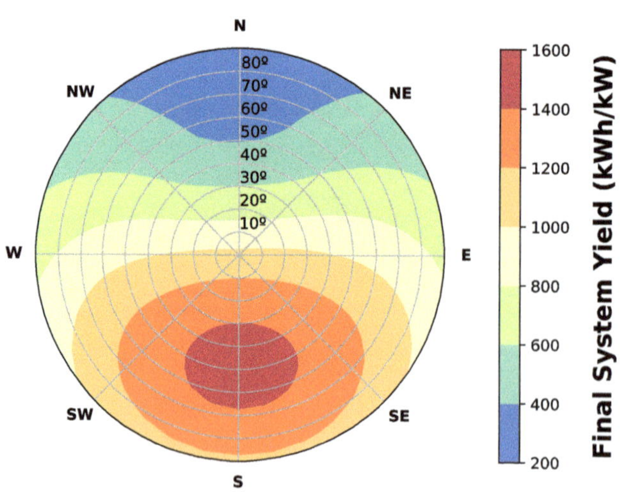

Figure 5.71 PV potential heatmap for Camrose, Canada. Credits: Ana Marcos Castro, CIEMAT.

Figure 5.67 Pedestrian view of Jeanne & Peter Lougheed Performing Arts Centre. Credits: Gordon Howell, HME.

Figure 5.68 Detail of the BIPV rainscreen at Jeanne & Peter Lougheed Performing Arts Centre. Credits: Gordon Howell, HME.

Figure 5.69 Rooftop string inverters at Jeanne & Peter Lougheed Performing Arts Centre. Credits: Credits: Gordon Howell, HME.

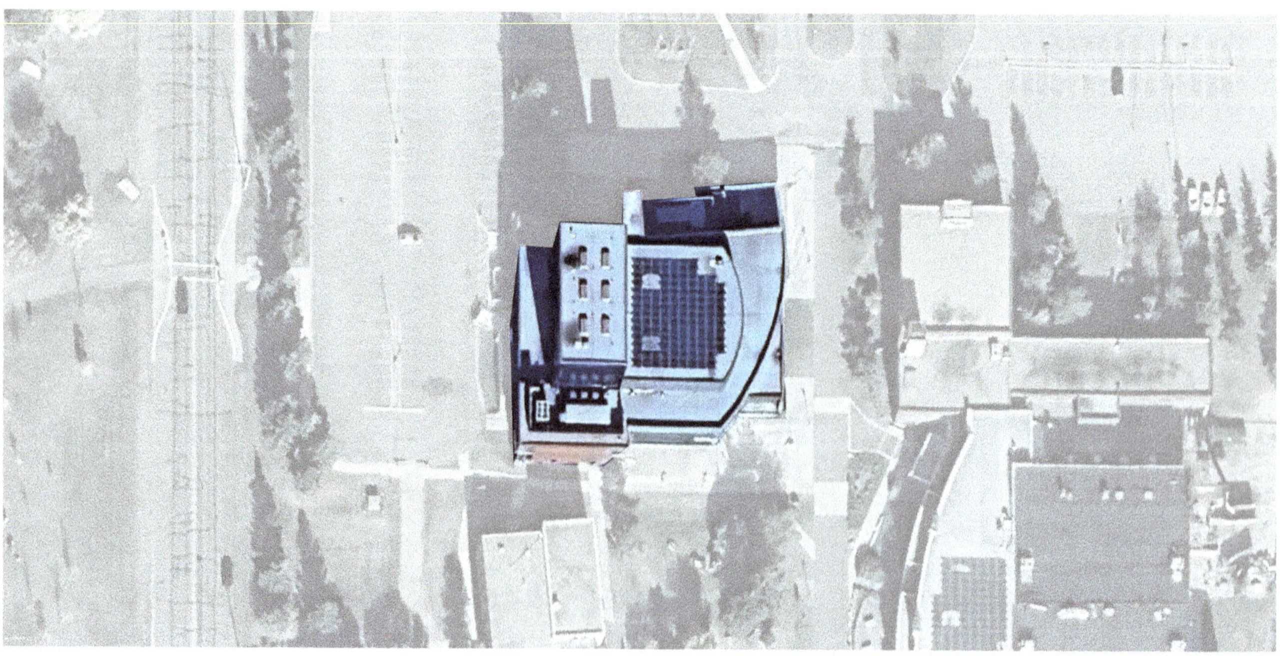

Figure 5.70 Site plan of Jeanne & Peter Lougheed Performing Arts Centre. Source: Google Maps.

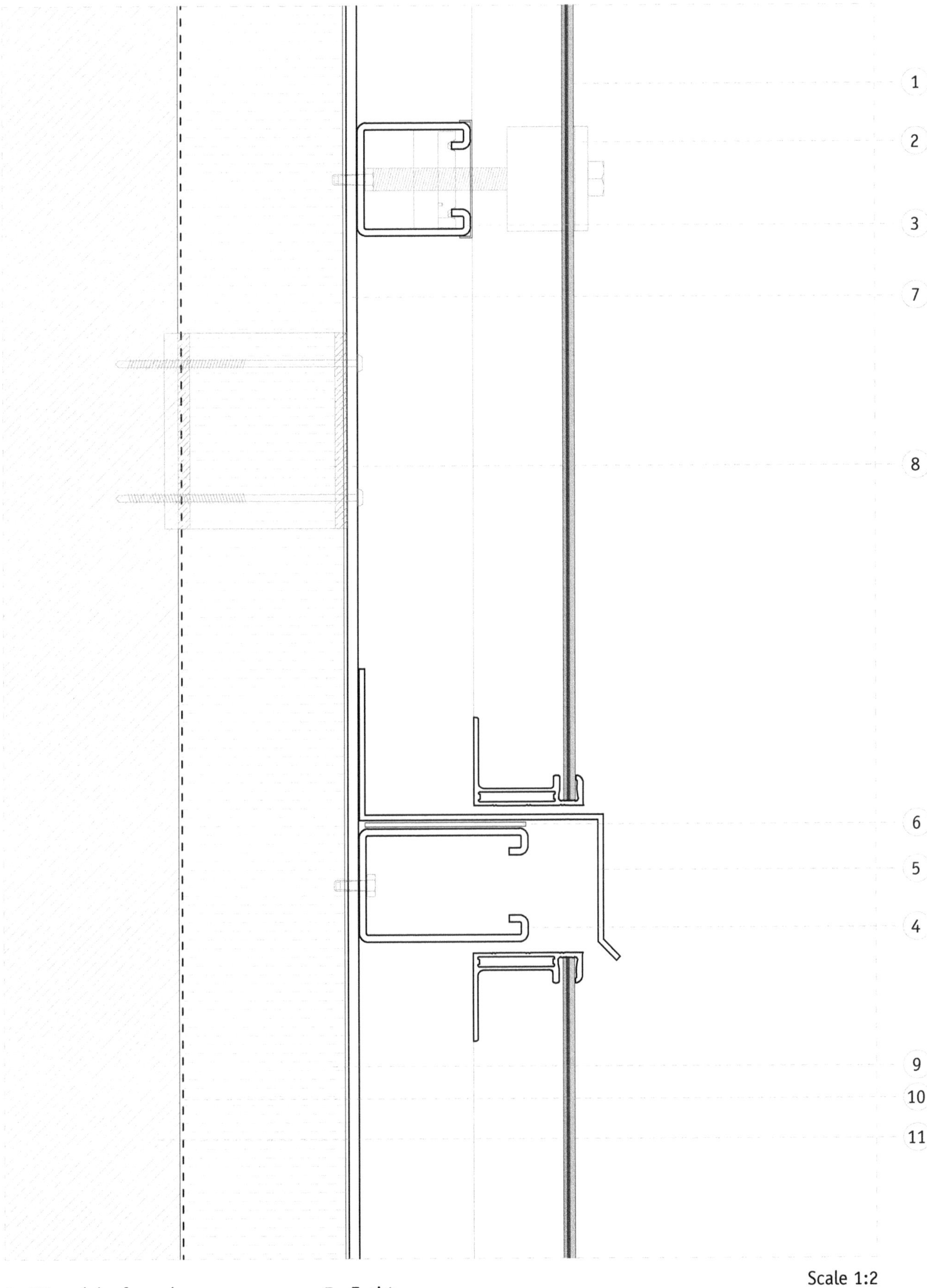

1. PV module, framed
2. Mid-clamp
3. Clamping strut
4. Transom
5. Inter-row flashing
6. Anti-corrosion tape
7. Z-girt
8. Wall bracket, thermally broken
9. Insulation
10. Water and air barrier
11. Primary structure

Figure 5.72 Jeanne & Peter Lougheed Performing Arts Centre. Cross-section drawing of the BIPV rainscreen installation Credits: adapted by Ana Marcos Castro from Howell-Mayhemm Engineering ©.

Tuas Port administrative building

LOCATION	Tuas, Singapore (1° 14′ 24″ N)
USE	Office
CLIENT	PSA Corporation
PV PRODUCER	Kromatix SA (mounting system by SolarGy)
ARCHITECT	ID Architects
ENGINEER	PDC Consulting Engineers
INSTALLER	Chan Rong Fen Building Construction Pte Ltd and SolarGy
YEAR OF CONSTRUCTION	2021

Project description: The PSA administrative building was built in 2021 and is located at the Tuas Mega Port in Singapore. The six-storey building is part of the Tuas Port Maintenance complex and received the Green Mark Award (Platinum) under the Super Low Energy Building (SLEB) category from Singapore's Building and Construction Authority for its best-in-class energy performance, use of renewable energy, and intelligent energy management strategies. It is a net-zero energy building thanks to its low-energy consumption (58% less energy than similar-sized buildings) offset by the PV electricity produced by a solar façade and a conventional PV rooftop system.

BIPV system key characteristics: The 961 BIPV modules cover 1,585 m² of the south, east, and west façades of the building. Free from any shade, the 250 kW BAPV façade is split into four arrays connected to four 60 kW grid-connected inverters. Each array consists of 11 strings. An additional PV rooftop system assists with the balance between energy demand and on-site energy generation. The 1674 mm × 985 mm × 8.5 mm monocrystalline silicon modules have been designed with architectural integration and energy performance in mind. Their frontsheet uses anthracite-coloured tempered glass with anti-reflective coating while the backsheet is heat-strengthened glass.

1.7 MWh/m²
Annual horizontal solar irradiation

Tropical Rainforest – Af
Köppen climate classification

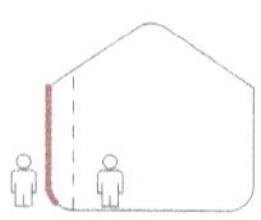

Rainscreen
BIPV application

New
Type of construction

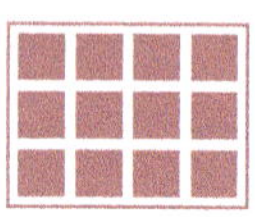

Opaque m-Si PERC
PV technology

249.86 kW
BIPV array nominal power

84.7 kWh/m²
Annual electricity generation per BIPV module area

134.7 MWh
Annual electricity generation

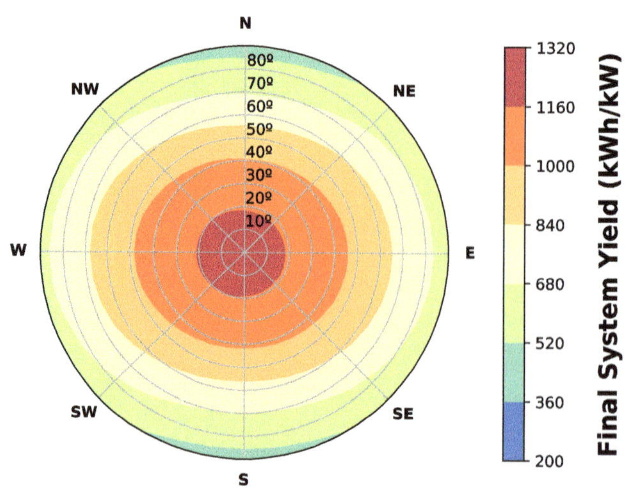

Figure 5.76 PV potential heatmap for Tuas, Singapore. Credits: Ana Marcos Castro, CIEMAT.

Figure 5.73 Pedestrian view of the Tuas Port administrative building. Credits: PSA.

Figure 5.74 Eagle-eye view of the Tuas Port building complex. Credits: PSA.

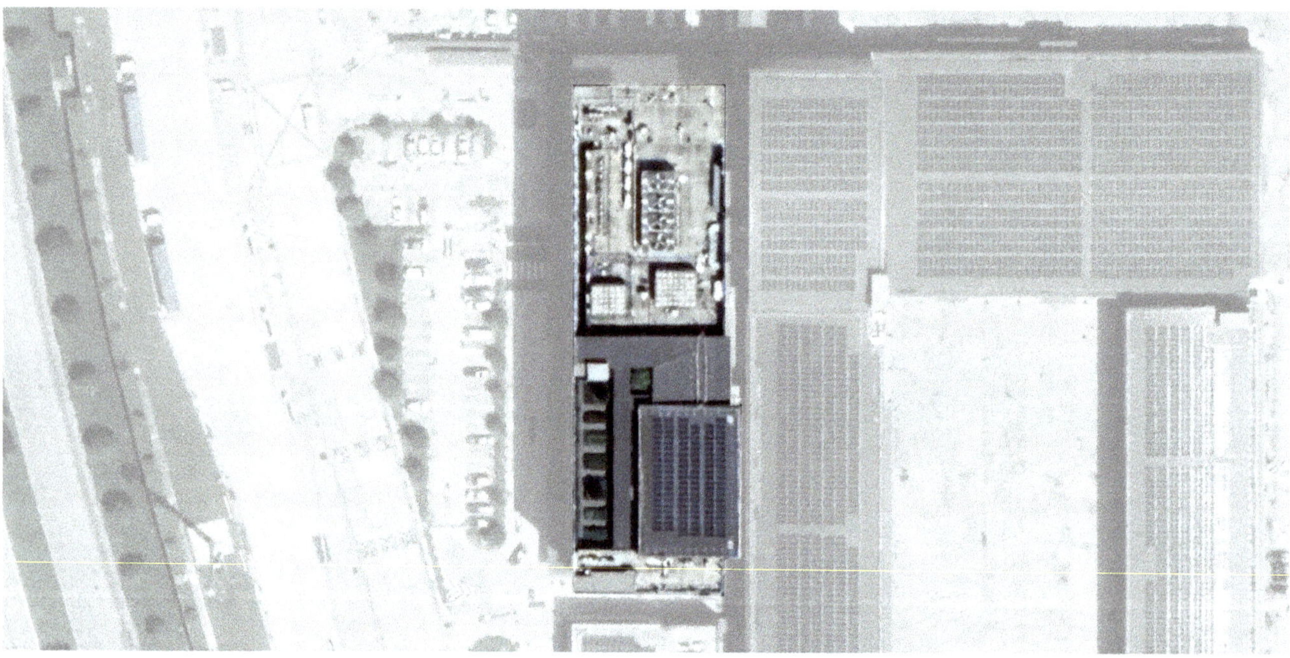

Figure 5.75 Site plan of the Tuas Port administrative building. Source: Google Maps.

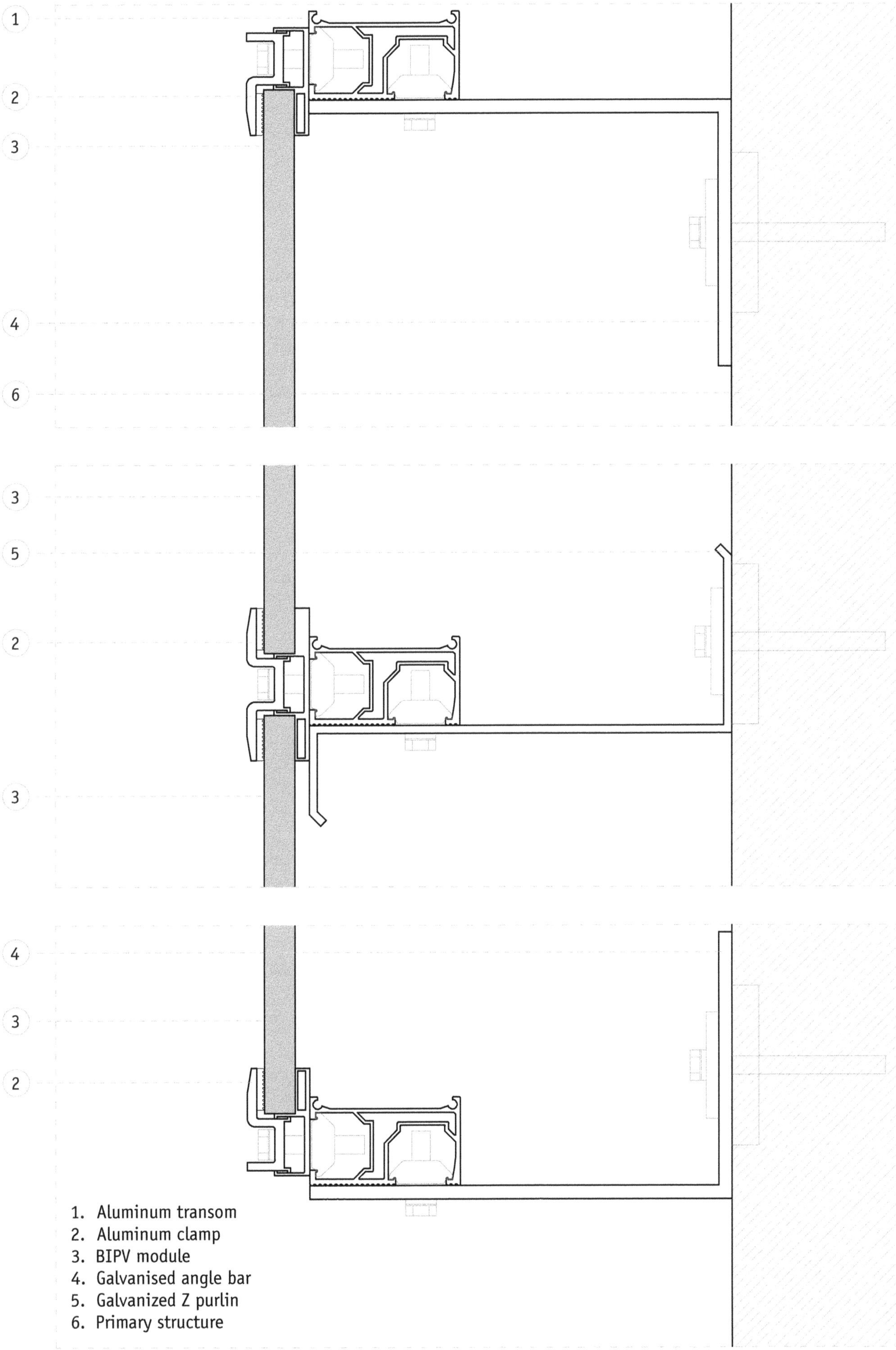

Scale 1:2

Figure 5.77 Tuas Port administrative building. Cross-section drawing of the BIPV rainscreen installation. Credits: adapted by Ana Marcos Castro from ID Architects PDE ©.

DeltaROSSO

LOCATION	**Vacallo, Switzerland (45° 50′ 32″ N)**
USE	**Residential & Office**
PV PRODUCER	**helioSKIN designed by deltaZERO, manufactured by ISSOL**
ARCHITECT	**deltaZERO- S. De Angelis & M. Mazza**
ENGINEER	**TermoConsult**
INSTALLER	**Greenkey Sagl**
YEAR OF CONSTRUCTION	**2016**

Project description: DeltaROSSO is a four-storey mixed-use building resting on an existing substructure of two basement floors. With a total of 1,651 m² of reconfigurable living and working space, DeltaROSSO meets carbon-neutral performance targets with EUI of 54 kWh/m²/year and TEDI of 5.1 kWh/m²/year. The building uses an air-to-water heat pump for space heating and cooling and an ERV. Apart from the use of BIPV modules, the façade and the roof also incorporate solar thermal panels. The BIPV and solar thermal elements are seamlessly integrated on the south-facing roof and façade for on-site electricity generation and domestic hot water heating, respectively.

BIPV system key characteristics: The cladding of DeltaROSSO is the first application example of "helioSKIN" (patented by deltaZERO), a building envelope cladding system that integrates energy production technologies. The building envelope consists of 203 black, frameless glass/glass BIPV modules and 13 solar thermal collectors integrated throughout the façade and roof. The roof includes 109 m² of BIPV tiles with a peak power of 17 kW, while the surface area on the BIPV façade is 194 m² of BIPV modules with a peak power of 30 kW. Both roof and façade systems are fastened to the building structure through a cold-formed galvanised steel railing system.

1.1 MWh/m²
Annual horizontal solar irradiation

Humid temperate – Cfb
Köppen climate classification

Roof and rainscreen
BIPV application

New
Type of construction

Opaque m-Si
PV technology

47 kW
BIPV array nominal power

161 kWh/m²
Annual electricity generation per BIPV module area

48.9 MWh
Annual electricity generation

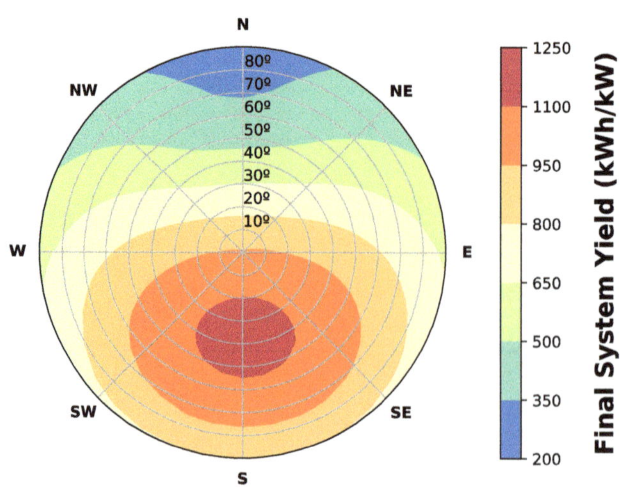

Figure 5.82 PV potential heatmap for Vacalla, Switzerland. Credits: Ana Marcos Castro, CIEMAT.

Figure 5.78 Pedestrian view of DeltaROSSO builing. Credits: Isa Zanetti, SUPSI.

Figure 5.79 Detail of the DeltaROSSO BIPV rainscreen. Credits: Isa Zanetti, SUPSI.

Figure 5.80 DeltaROSSO BIPV ventilated roof under construction. Credits: Isa Zanetti, SUPSI.

Figure 5.81 Site plan of DeltaROSSO. Source: Google Maps.

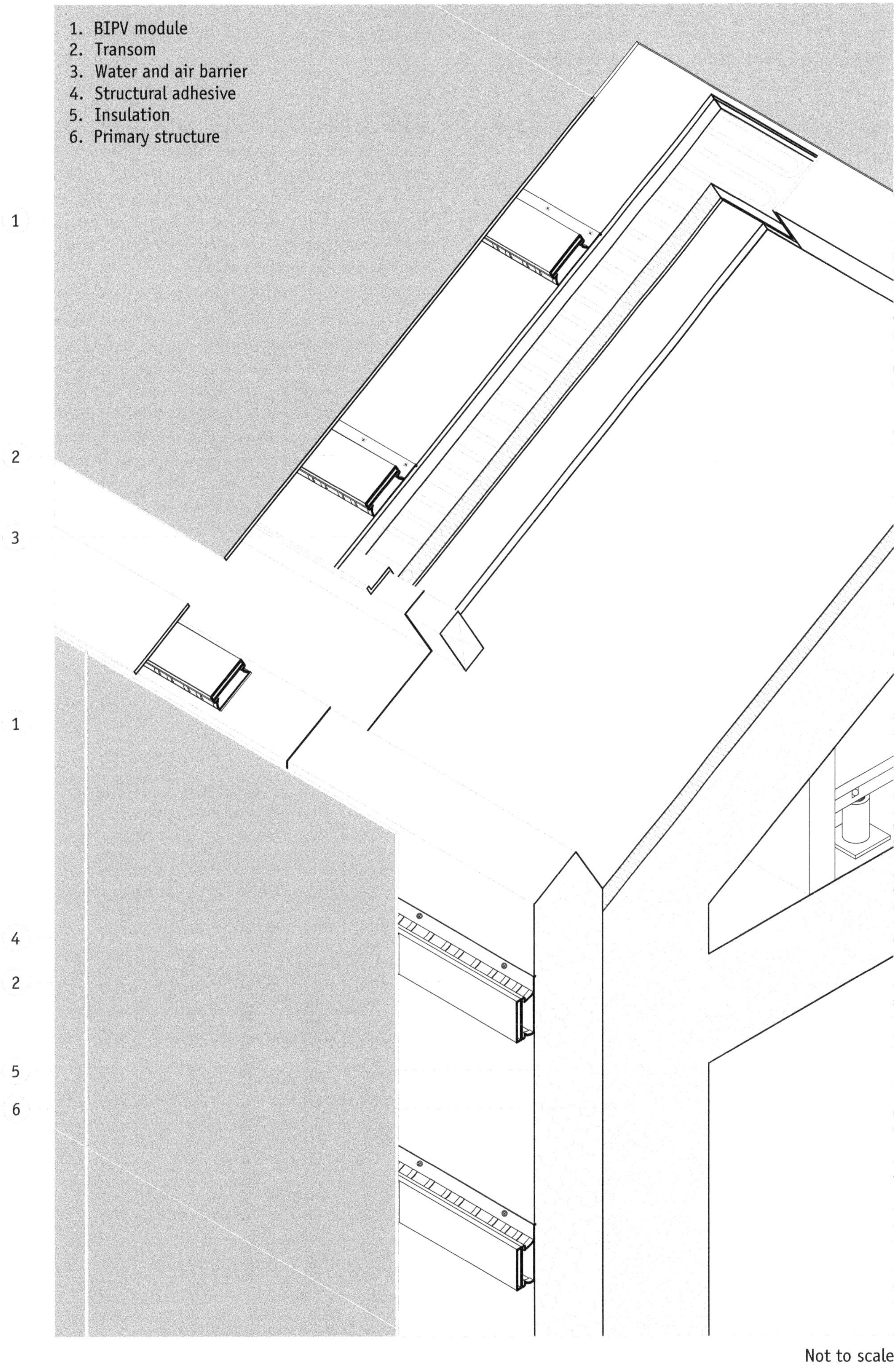

Figure 5.83 DeltaROSSO. Axonometric drawing of the BIPV rainscreen installation. Credits: adapted by Ana Marcos Castro from deltaZERO ©. Credits: S. De Angelis & M. Mazza, deltaZERO.

Manitou a Bi Bii Daziigae University Building

LOCATION	**Winnipeg, Manitoba (49° 53′4″ N)**
USE	**Academic**
PV PRODUCER	**SolarLab**
ARCHITECT	**Diamond Schmitt Architects and Number TEN Architectural Group**
ENGINEER	**SolarLab, SMS Engineering, Crosier Kilgour**
INSTALLER	**Flynn**
YEAR OF CONSTRUCTION	**2022**

Project description: The Manitou a Bi Bii Daziigae ("where Creator sits and Brings light") building at Red River College Polytechnic brings together students with education and industry professionals in new ways that facilitate social innovation, enterprise, and innovative research. The 9,300 m² facility unites a repurposed concrete framed heritage building and new construction to create an engaging crossroads in Winnipeg's historic Exchange District. The new building addresses the College's Energy Use Intensity (EUI) target of 100 kWh/m²/year through a high-performance BIPV envelope, energy efficiency measures, and a PV rooftop system. Together, the two buildings house a unique combination of indigenous and international student spaces.

BIPV system key characteristics: The 710 m² BIPV rainscreen is the first installation of its kind in Canada, developed through an integrated design process that included extensive coordination, modelling, testing, and mock-ups. The frontsheet of the monocrystalline BIPV modules is made of a physical vapor deposition (PVD) coated glass that changes colour depending on the view angle and the weather. A conventional rooftop array adds about 800 m² of bifacial PV modules. In the first year of use, the two systems produced as much as 70%–80% of the daily building electricity requirements during the shoulder seasons.

1.37 MWh/m²
Annual horizontal solar irradiation

Humid continental – Dfb
Köppen climate classification

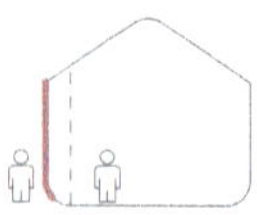

Rainscreen
BIPV application

New
Construction

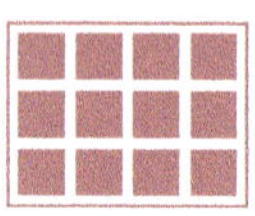

Opaque coloured m-Si (BIPV), Bifacial m-Si (rooftop)
PV technology

94 kW (BIPV), 54 kW (rooftop)
BIPV array nominal power

112 kW/m²
Annual electricity generation per BIPV module area

80 MWh (BIPV) + 89 MWh (rooftop)
Annual electricity generation

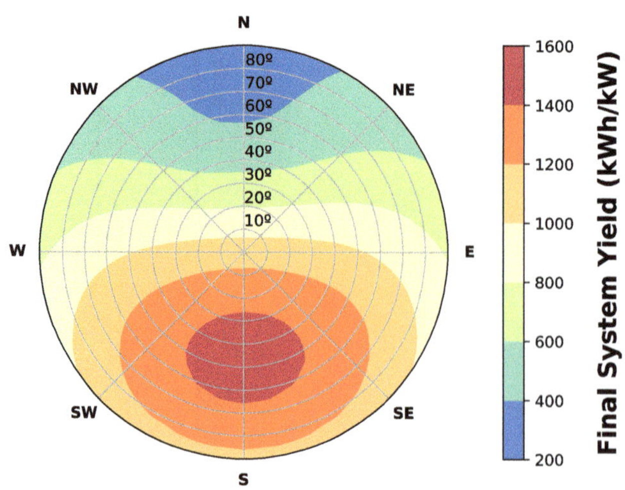

Figure 5.87 PV potential heatmap for Winnipeg, Canada. Credits: Ana Marcos Castro, CIEMAT.

Figure 5.84 Pedestrian view of the Manitou A Bi Bii Daziigae University building. Credits: Diamond Schmitt Architects.

Figure 5.85 BIPV rainscreen detail of Manitou A Bi Bii Daziigae University building. Credits: Diamond Schmitt Architects.

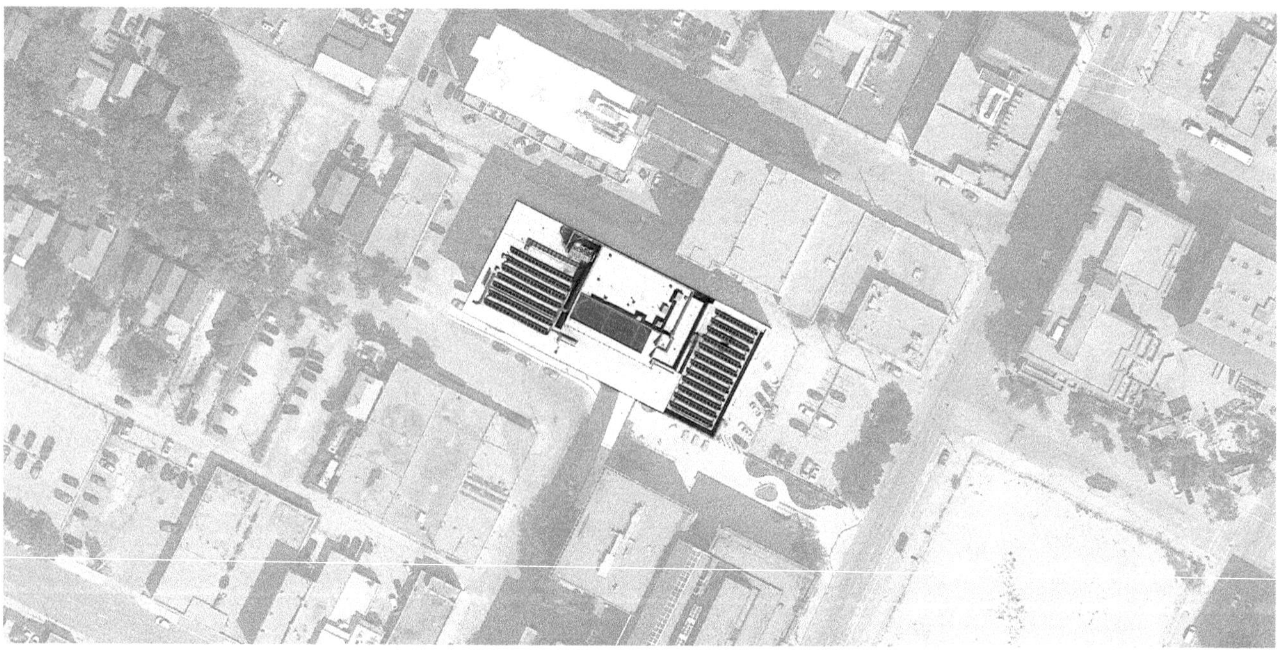

Figure 5.86 Site plan of the Manitou A Bi Bii Daziigae University building. Source: Google Maps.

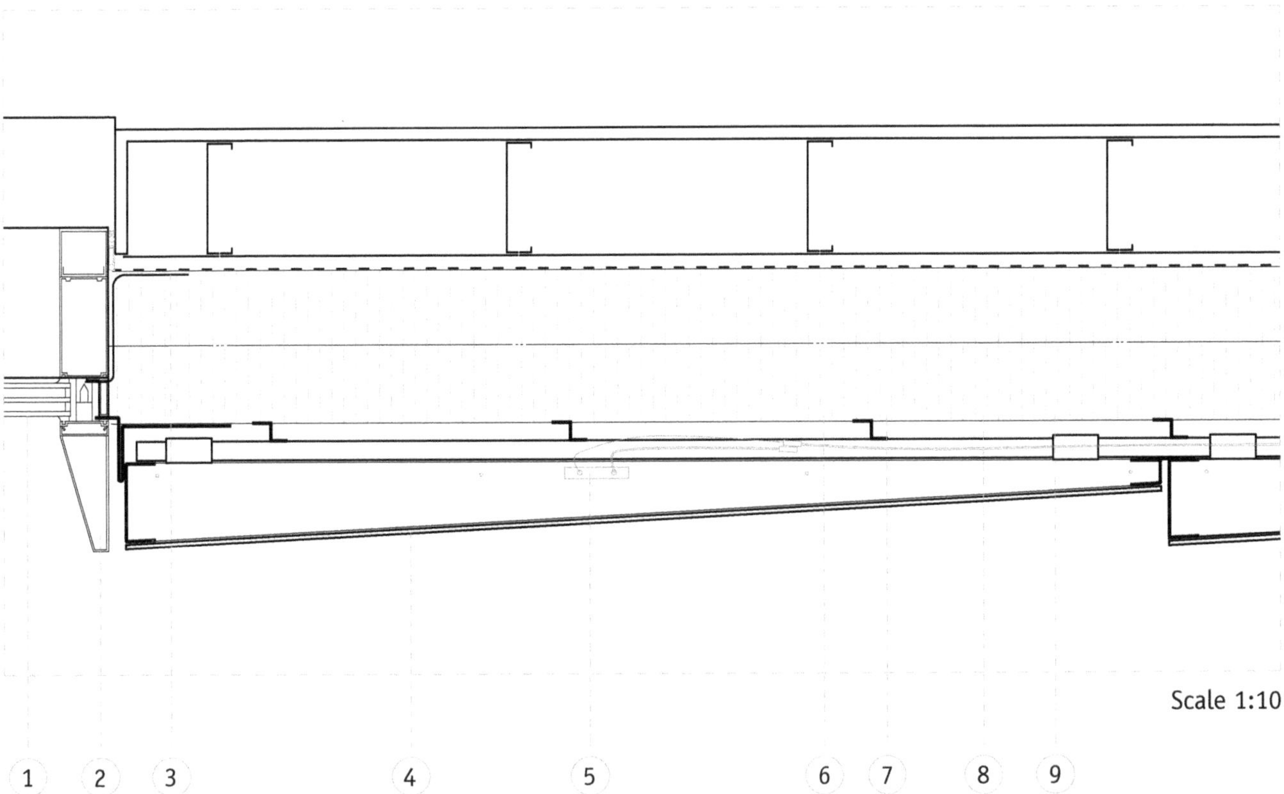

1. Triple glazed IGU
2. Curtain wall pressure plate
3. Silicone transition membrane
4. BIPV module and cassette
5. BIPV junction box and optimizer
6. Fiberglass thermal clip
7. Mineral wool insulation
8. Water barrier
9. Air and vapour barrier

Figure 5.88 Manitou A Bi Bii Dazigae University building. Cross-section drawing of the BIPV rainscreen installation. Source: adapted by Ana Marcos Castro from Diamond Schmitt Architects ©.

Fanshawe College Innovation Village

LOCATION	London, Ontario (42° 59′ 1″ N)
USE	Research Facility
PV PRODUCER	SolarLab
ARCHITECT	Diamond Schmitt Architects
ENGINEER	SolarLab, German Solar & Smith +Andersen
INSTALLER	German Solar
YEAR OF CONSTRUCTION	2023

Project description: The 11,900 m² Innovation Village is a vibrant and collaborative space where students develop entrepreneurial skills, learn to create start-up businesses, and establish industry partnerships. Constructed at the centre of Fanshawe College's main campus, the Innovation Village brings together students, professors, and professionals from all faculties to engage in cross-disciplinary collaboration to advance ideas, test products, and provide opportunities for new concepts. The building's BIPV cladding aligns with this innovation mandate and comes online as the College completes a new Campus Energy Centre.

BIPV system key characteristics: The cladding includes custom-sized BIPV and dummy modules using physical vapor deposition (PVD) coated glass. The BIPV modules range from 120–470 W. Early cost-benefit analysis demonstrated the positive return of investment (ROI) for BIPV façade has been installed on all sides of the new tower, including the north-facing elevation. The cladding is composed of alternating upward-tilted BIPV modules and smaller off-set dummy panels, creating a distinctive form for the new structure. This custom cladding is estimated to provide approximately 18% of the project's near-term power requirements. When the upper floors of the project are fitted out and the planned future rooftop PV array is installed, 30% of the total power demand will be generated by the combined BIPV and rooftop systems.

1.5 MWh/m²
Annual horizontal solar irradiation

Humid continental- Dfb
Köppen climate classification

Rainscreen
BIPV application

New
Construction

Opaque coloured m-Si
PV technology

174.9 kW
BIPV array nominal power

263 kWh/m² (for active cell area only)
Annual electricity generation per BIPV module area

123 MWh (façade only)
Annual electricity generation

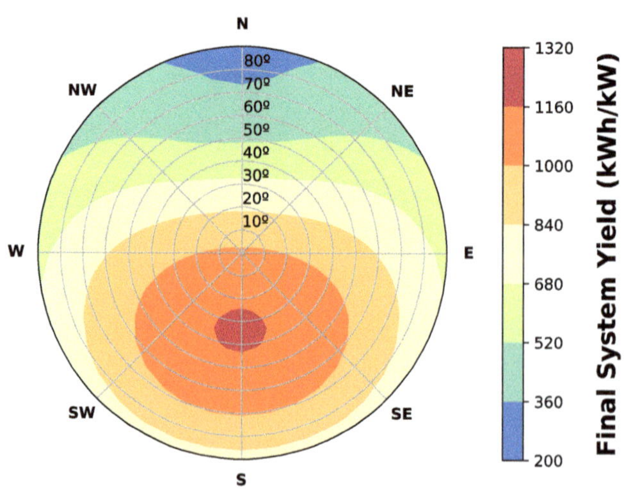

Figure 5.93 PV potential heatmap for London, Canada. Credits: Ana Marcos Castro, CIEMAT.

Figure 5.89 Pedestrian view of Fanshawe College Innovation Village. Credits: Diamond Schmitt Architects.

Figure 5.90 Close-up of the Fanshawe College Innovation Village BIPV rainscreen. Credits: Diamond Schmitt Architects.

Figure 5.91 Detail of the BIPV rainscreen. Credits: Diamond Schmitt Architects.

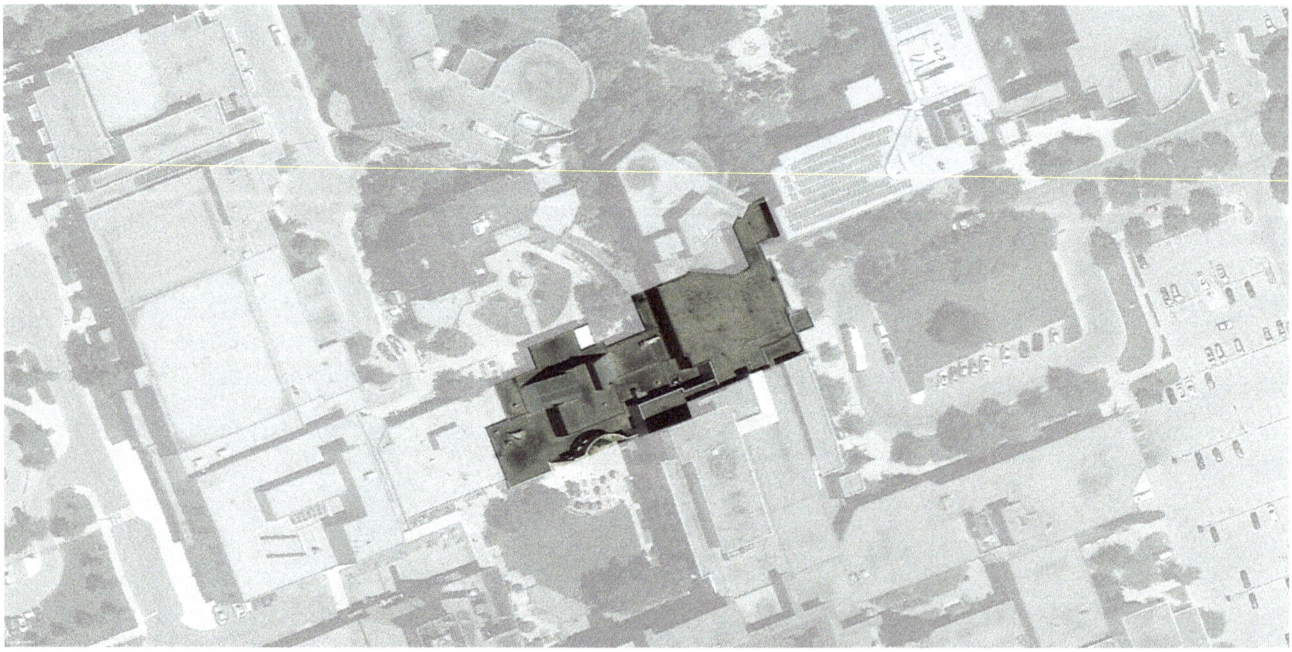

Figure 5.92 Site plan of Fanshawe College Innovation Village. Source: Google Maps.

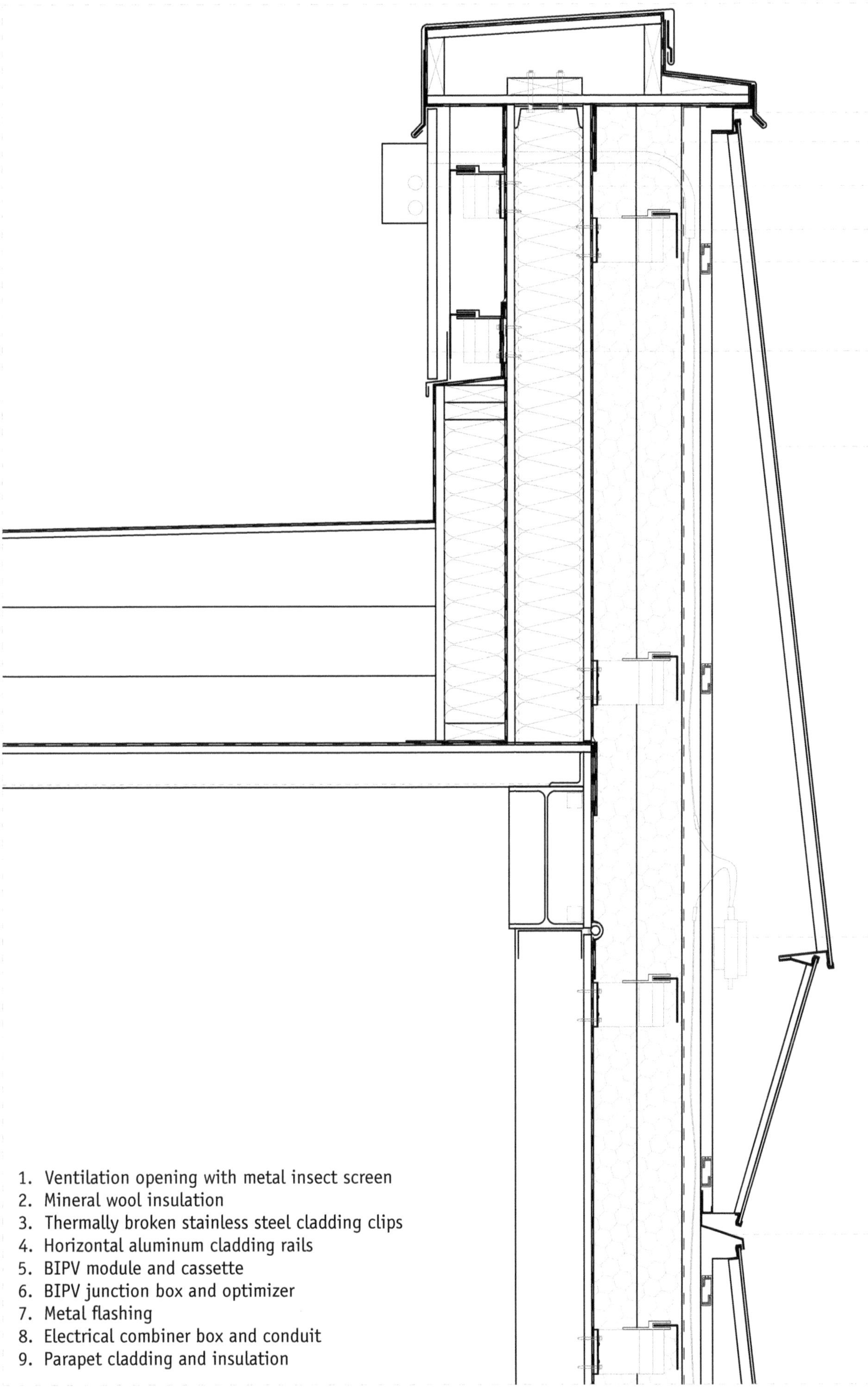

Scale 1:12

Figure 5.94 Fanshawe College Innovation Village. Cross-section drawing of the BIPV rainscreen installation. Credits: adapted by Ana Marcos Castro from Diamond Schmitt Architects ©.

Genyo Research facility

LOCATION	Granada, Spain (37° 8′ 56″ N)
USE	Research Facility
CLIENT	Pfizer and Andalusian Government
PV PRODUCER	Onyx Solar (mounting system by Jansen)
ARCHITECT	Enrique Vallecillos and Emiliano Rodríguez (PLANHO)
ENGINEER	Hinsa
INSTALLER	Onyx Solar
YEAR OF CONSTRUCTION	2010

Project description: Constructed in 2010, the building is located at the Health Science Technological Park (PTS) in Granada, Spain. It hosts a research centre on Genomics and Oncology supported by Pfizer company, the University of Granada, and the Andalusian Government. Genyo was designed as a hub for interdisciplinary research between health, university, and industry professionals. The BIPV double-skin façade, made of two different coloured BIPV modules, supplies the building with enough electricity to cover 100% of the centre's annual lighting loads while reducing solar heat gains and, thus, energy demand.

BIPV system key characteristics: The 550 m² BIPV double-skin façade is made up of two coloured amorphous silicon glass BIPV modules, randomly alternating between dark grey and blue. The BIPV modules consists of a triple laminate safety glass that has been developed specifically for the project and displays a visible transmittance of 20%. The colour is achieved through two screen-printed inactive glass laminates placed in between glass panes, making the photovoltaic film invisible.

1.7 MWh/m²
Annual horizontal solar irradiation

Hot-summer Mediterranean – Csa
Köppen climate classification

Double-skin façade
BIPV application

New
Type of construction

Semi-transparent amorphous-Si
PV technology

19.3 kW
BIPV array nominal power

58.2 kWh/m²
Annual electricity generation per BIPV module area

32 MWh
Annual electricity generation

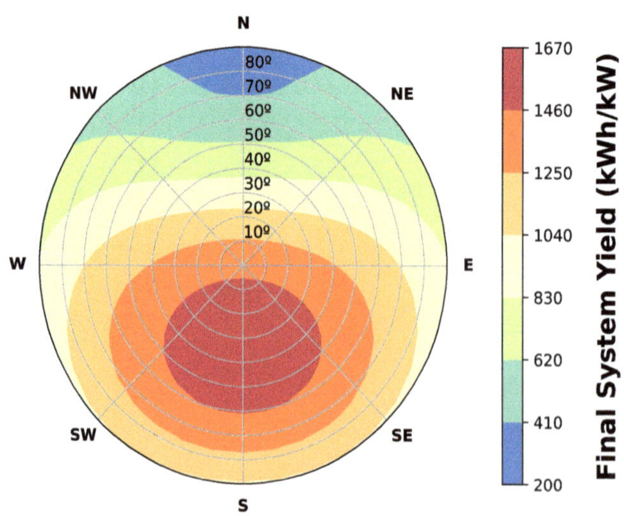

Figure 5.98 PV potential heatmap for Granada, Spain. Credits: Ana Marcos Castro, CIEMAT.

Figure 5.95 Pedestrian view of the Genyo Research facility, showing the BIPV south and west façades under construction. Credits: Onyx Solar.

Figure 5.96 Detail of the BIPV double skin façade of the Genyo Research facility. Credits: Onyx Solar.

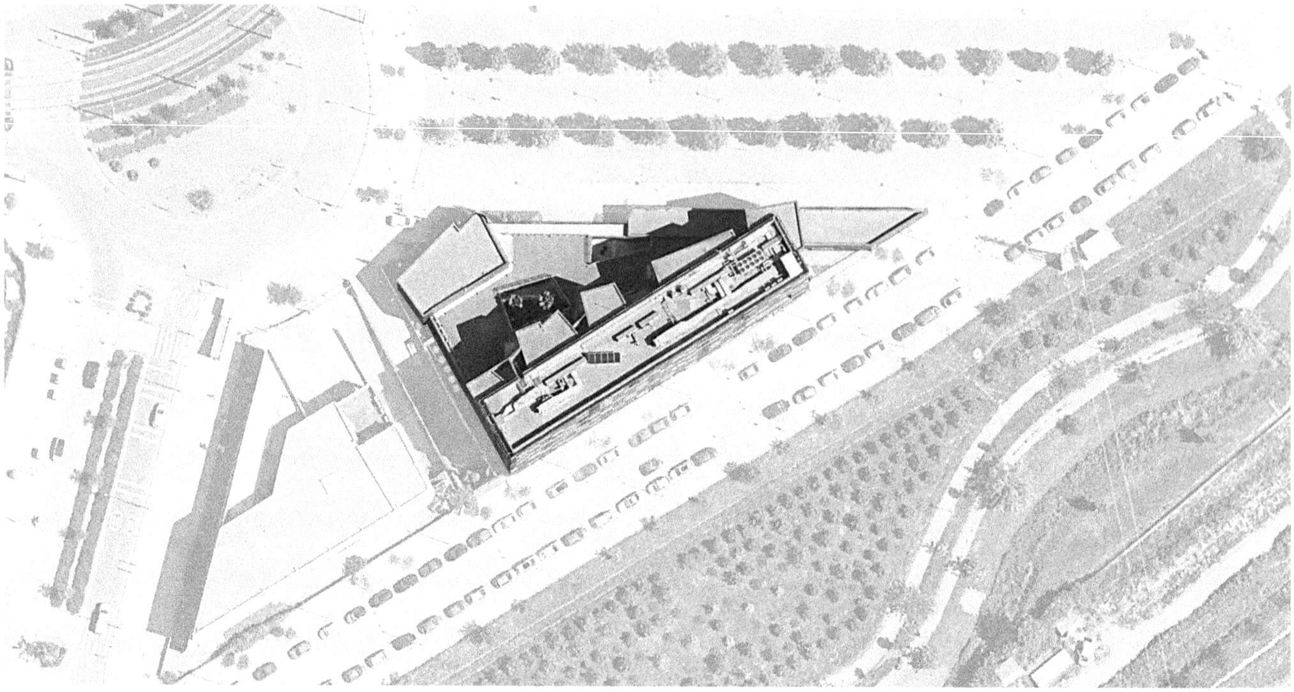

Figure 5.97 Site plan of Genyo Research facility. Source: Google Maps.

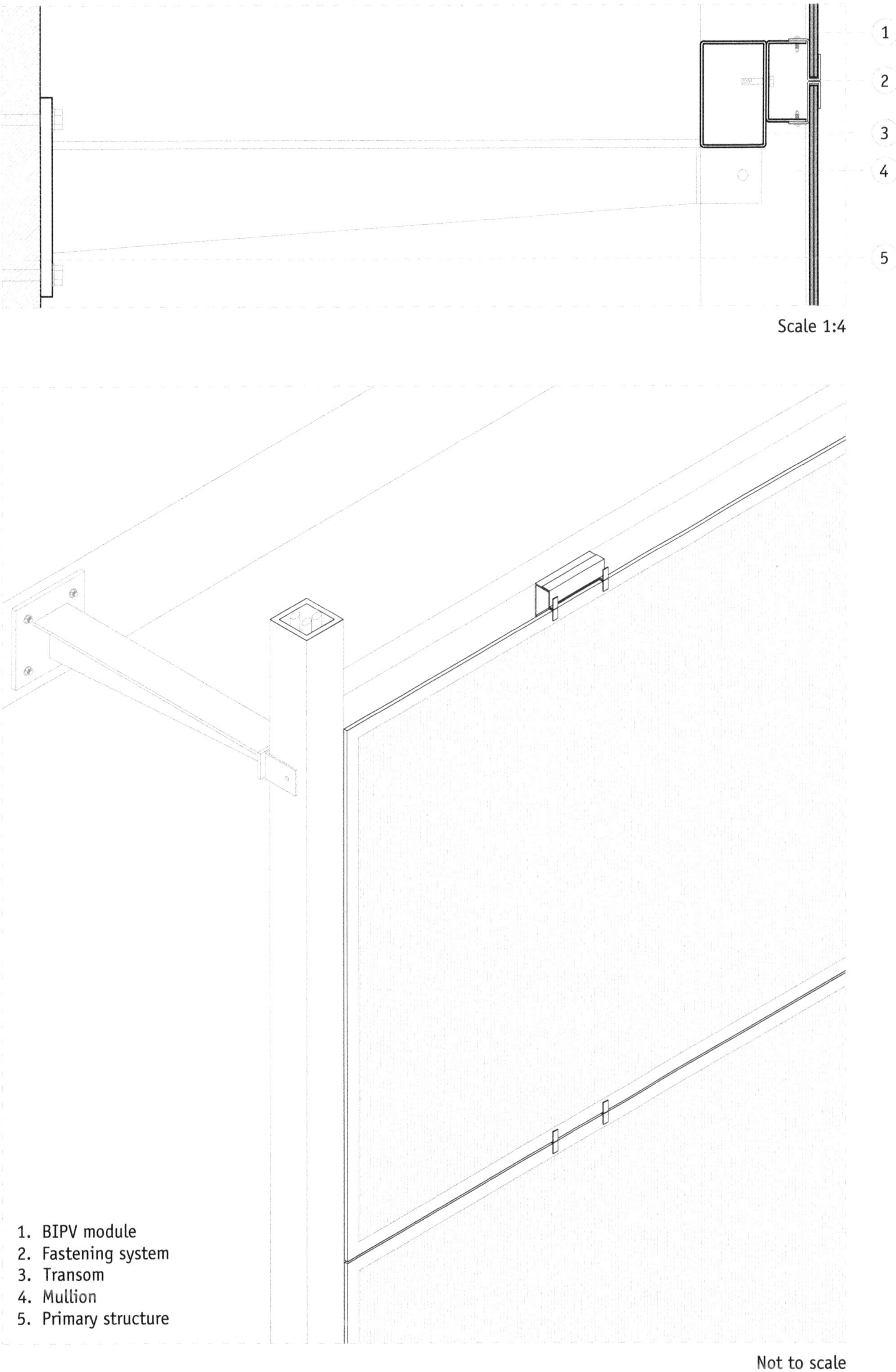

Figure 5.99 Genyo Research facility. Cross-section and axonometric drawings of the BIPV double skin façade. Credits: adapted by Ana Marcos Castro from Onyx Solar ©.

Palazzo Argonauta

LOCATION	Rome, Italy (41° 51′ 55″ N)
USE	Office
PV PRODUCER	Sunerg Solar Energy PV modules (100 mm × 150 mm)
ARCHITECT	Agenzia di Architettura
ENGINEER	Cool Projects S.r.l.
INSTALLER	Genera S.p.A.
YEAR OF RENOVATION	2019

Project description: One of the largest buildings in Rome, Palazzo Argonauta (70,000 m² and 275,000 m³), underwent a retrofit intervention completed in 2019 to incorporate PV into its south façade. The building is used for office space and has a total of 10 storeys. In addition to the façade renovation, the building systems (e.g., HVAC and lighting) were also renovated for efficiency while maintaining comfortable conditions for the 4,000 daily users.

BIPV system key characteristics: The biggest challenge was to make the system almost invisible, playing on the "mystification" of reality, i.e., transmitting the illusion that the photovoltaic panels were the windows, while the brise-soleil elements were perceived, on the contrary, as stringcourse elements; on the west façade, a similar system was created, with blue enamelled glass panels of the same colour and size as the photovoltaic ones, anchored with the same supports, to give visual continuity to the active and passive façade.

1.3 MWh/m²
Annual horizontal solar irradiation

Hot-summer Mediterranean – Csa
Köppen climate classification

Rainscreen
BIPV application

Retrofit
Type of construction

Opaque mc-Si (regular PV module except for size)
PV technology

145 kW
BIPV array nominal power

140 kWh/m²
Annual electricity generation per BIPV module area

137.6 MWh
Annual electricity generation

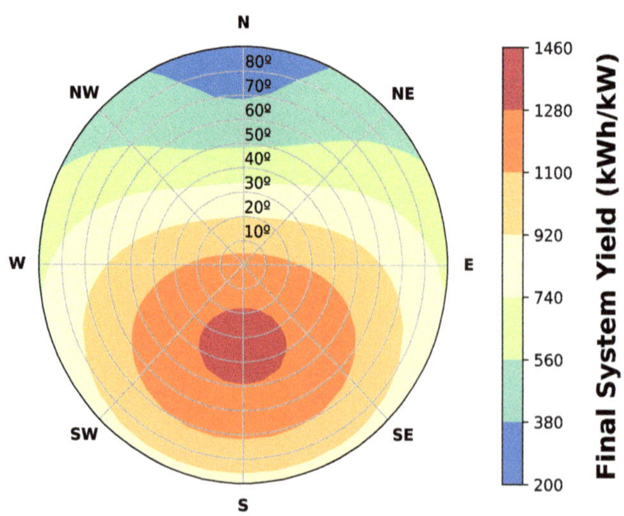

Figure 5.104 PV potential heatmap for Rome, Italy. Credits: Ana Marcos Castro, CIEMAT.

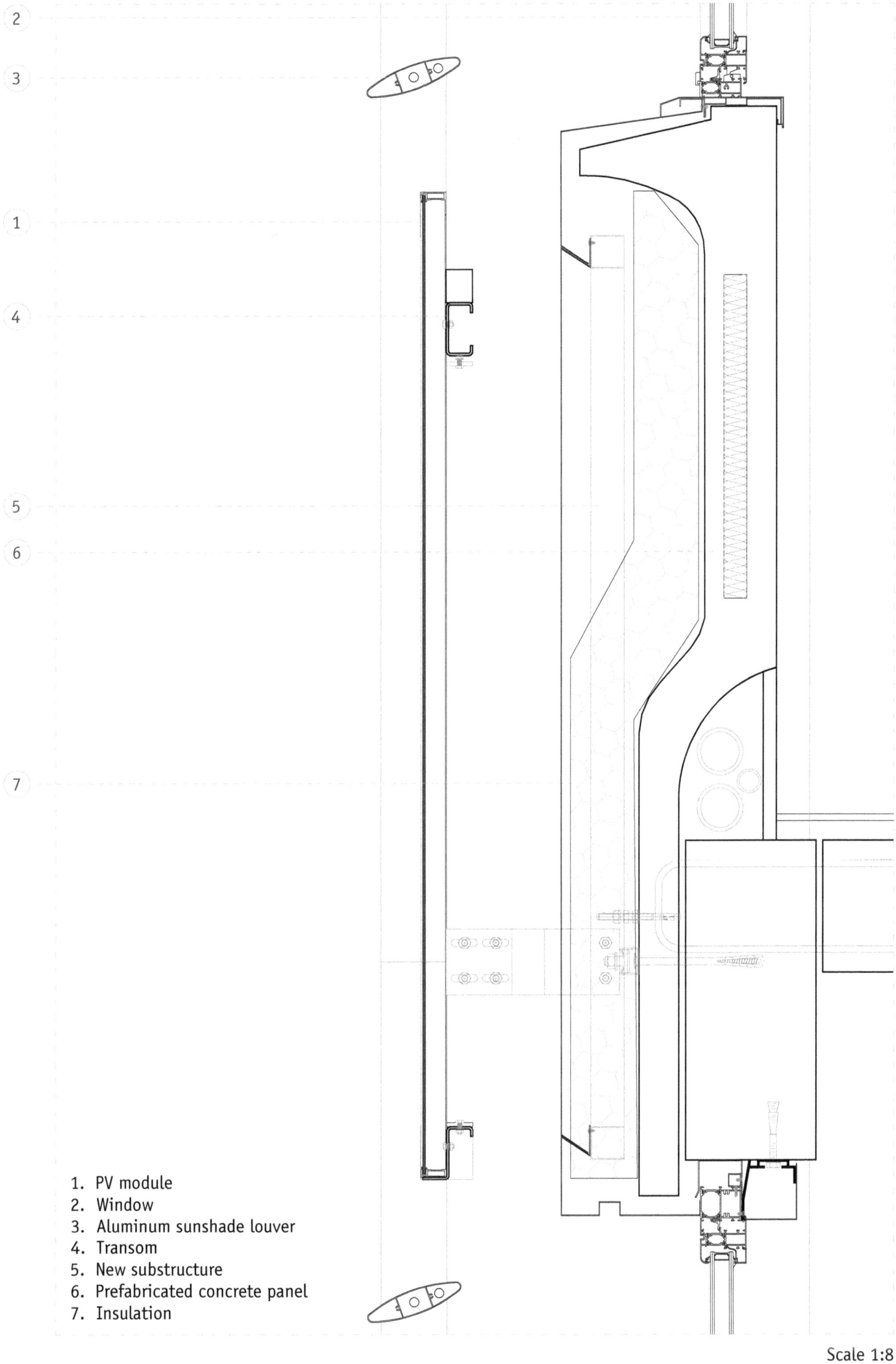

Scale 1:8

Figure 5.105 Palazzo Argonauta. Cross-section drawing of the BIPV double skin façade. Credits: adapted by Ana Marcos Castro from Agenzia di Architettura ©.

5.3.2 Curtain wall

General: This section reviews both stick and unitised curtain wall systems, with a focus on pressure-equalised configurations [9]. The following considerations refer to both the transparent (or semi-transparent) vision section and the opaque spandrel of a curtain wall.

Structural Design: BIPV curtain walls are non-load-bearing envelope structures that are designed to hang from building floor slabs ① or the load-bearing building structure, through steel bolted anchors. They should be engineered to withstand and transfer dead, environmental (e.g., wind, and occasionally snow and ice), live and impact loads, in compliance with the local building codes. The anchors ② should be installed to allow differential movement between the curtain wall and the building structure. The mullions ③ and transoms ④ should be sized to support the weight of multi-pane insulated glazing units (IGU) – including semi-transparent BIPV IGUs – installed in the vision sections ⑤ of the curtain wall, the single-pane laminated opaque BIPV glass installed in the spandrel sections ⑥, and any additional cladding elements such as solar shading. The use of heat-strengthened or fully tempered innermost glass is commonly used to enhance user safety from glass breakage. The BIPV glazing units are held in place using (1) mechanical-fastening pressure plates ⑦, (2) structural silicone or (3) toggles. Both stick and unitised systems should be designed to accommodate the thermal movement between the frame connections and between the frame and IGUs, including vertical framing deflection. Depending on the building use (e.g., government buildings) and location (e.g., hurricane or tornado prone region), blast-resistant BIPV glass products might be required.

Hygrothermal Performance: To prevent water penetration and ensure effective drainage, a pressure-equalised cavity is engineered within the curtain wall framing. This cavity, positioned between the inner gaskets or wet seals (serving as air barriers) and the exposed outer gasket or seal, effectively manages any potential moisture ingress. Importantly, the principle of a pressure-equalised cavity ⑧ can be applied regardless of the attachment method for the IGU, and it is suitable for both stick and unitised curtain wall systems. For durability and system longevity, the use of UV-resistant structural silicone or gaskets covered by low-profile protective caps ⑨ is advised. The gaskets and structural sealants should be inspected every five to ten years and replaced if needed. Any penetrations for wiring management through the framing should be engineered to maintain proper air and water tightness. The use of thermally broken frame should be favoured for system longevity and thermal performance. In cold climates, the use of thermally broken frame has the additional benefit of higher condensation resistance. The choice of thermal insulation material and thickness in the spandrel section depends on shadow box depth, local building codes requirements, fire safety regulations, and sustainability needs.

Fire Safety: The use of non-combustible insulation (e.g., mineral wool) is recommended ⑩ to reduce or prevent fire propagation through the spandrel section. The use of a fire barrier and a smoke seal ⑪ between the floor slab and the curtain wall is recommended as an effective fire and smoke control practice between floors.

Electrical Considerations: Similarly to skylight applications, DC wiring can be routed along the pressure-equalised cavity of the mullions and transoms. At the vision section of the curtain wall, the BIPV IGU should incorporate edge junction box ⑫ for ease of installation and maintenance.

Architectural Considerations: The selection of PV technology, colours, and transparency levels enables building designers to deliver an engaging, well-lit indoor environment, elevate the user's experience, and turn the BIPV curtain wall into a prominent architectural element. The installation of Si-based semi-transparent IGUs can create an interesting luminous indoor environment. At the same time, it will function as a shading device reducing solar heat gains and, thus, directly impacting building energy performance (heating, cooling, lighting, and solar electricity generation) and occupant comfort (visual and thermal). To optimise the design of a BIPV curtain wall system, the use of building performance simulation analysis is essential. The use of coloured BIPV modules on the spandrel can enrich the aesthetics of the building exterior. The use of BIPV laminated glass in the spandrel or as the outermost layer of the IGU in the vision section can also enhance acoustic performance over standard IGUs, due to the noise-attenuating properties of the BIPV laminate (i.e., PVB).

Sustainability: An optimally designed BIPV curtain wall system serves as a multifaceted solution towards sustainable and energy-efficient buildings, through the generation of renewable electricity and control of solar gains and daylight. A BIPV curtain wall system with moderate window-to-wall ratio of 50% or less can enhance solar energy generation and building efficiency without compromising occupant comfort [10]. When paired with high-performance curtain wall framing, like thermally broken mass timber, it facilitates buildings in progressing towards operational carbon neutrality and a sustainable future.

An axonometric of a typical BIPV stick curtain wall. The drawings illustrate all key components of a BIPV curtain wall system.

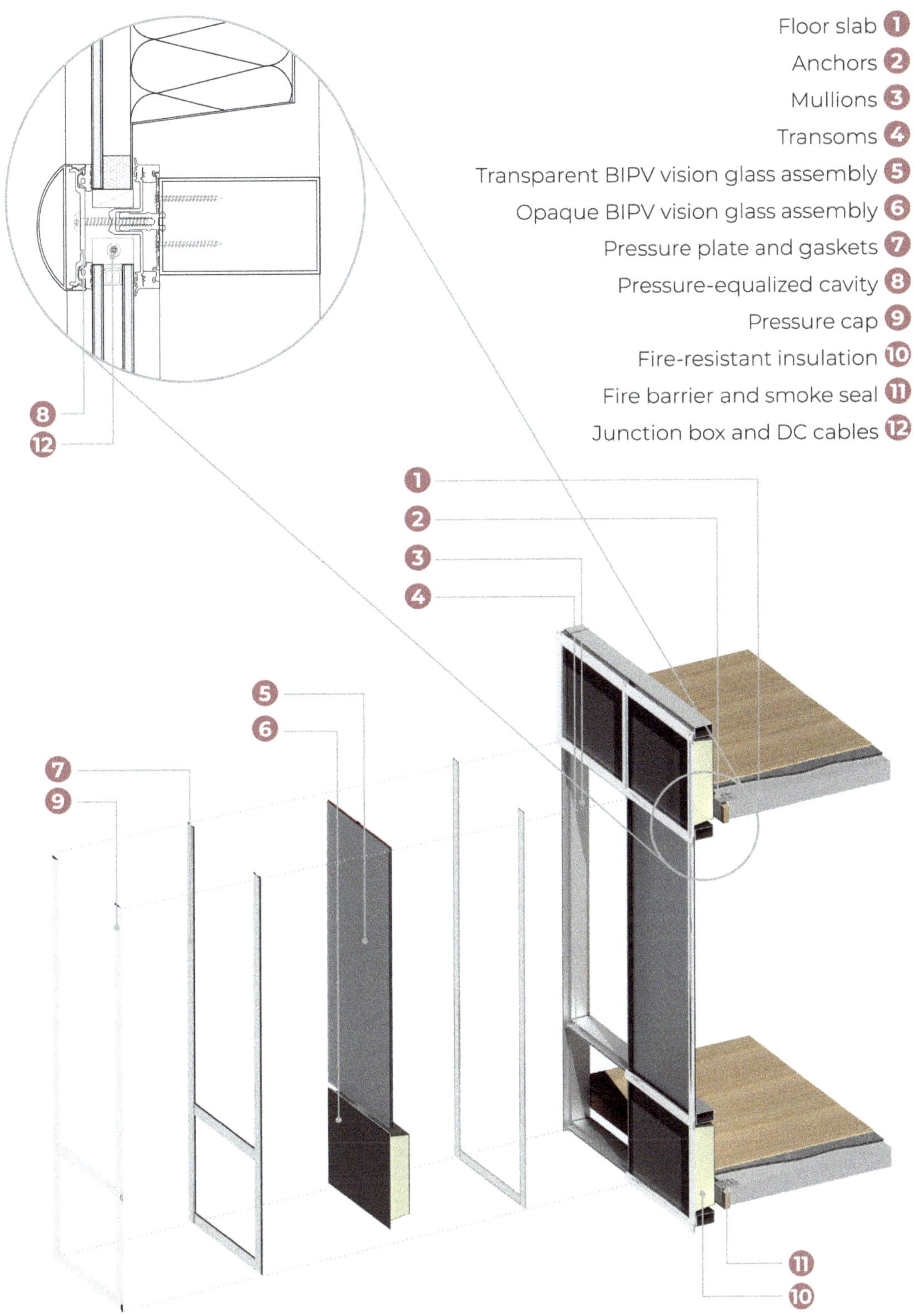

Figure 5.106 Exploded axonometric drawing of a typical BIPV curtain wall. Credits: Marley Dowling, Sara Eskandar, Amy Markwart.

Yanmar Tokyo building

LOCATION	Tokyo, Japan (35° 40′ 48″ N)
USE	Office & Retail
OWNER	Seirei Kosan Co
PV PRODUCER	AGC
ARCHITECT	Nikken Sekkei
ENGINEER	Takenaka Corporation (General Contractor) LIXIL Corporation (Façade Contractor)
INSTALLER	LIXIL Corporation (Unitised curtainwall)
YEAR OF CONSTRUCTION	2023

Project description: Centrally located in front of the Yaesu exit of Tokyo station, this new 14-storey-high and 22,000 m² complex houses the new offices for Yanmar Holdings. The building is part of the redevelopment of the Chiyoda City and is directly connected to Tokyo's Central Station and the Tokyo Midtown Yaesu shopping centre. In addition to offices space, the building also includes commercial and retail space, galleries, and event spaces.

BIPV system key characteristics: The BIPV modules are incorporated on the upper section of the curtain wall and the fly-by parapet of the southeast-facing façade of the building. With a total surface area of 77.5 m², the BIPV system incorporates four different module sizes, customised to match the linear façade design: 615 mm × 2360 mm, 615 mm × 2040 mm, 312 mm × 2360 mm, and 312 mm × 2040 mm. Besides electricity generation, the BIPV parapet has the functional role of hiding the view of rooftop mechanical equipment units from the ground level.

1.5 MWh/m²
Annual horizontal solar irradiation

Humid subtropical – Cfa
Köppen climate classification

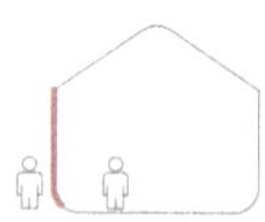

Curtain wall
BIPV application

New
Construction

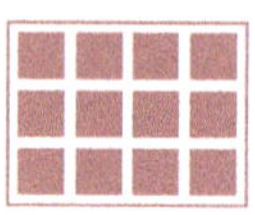

Opaque m-Si
PV technology

9.5 kW
BIPV array nominal power

73.9 kWh/m²
Annual electricity generation per BIPV module area

5.7 MWh
Annual electricity generation

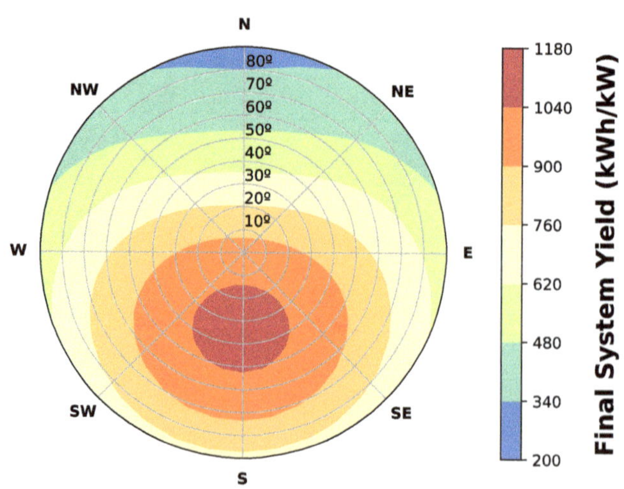

Figure 5.110 PV potential heatmap for Tokyo, Japan Credits: Ana Marcos Castro, CIEMAT.

Figure 5.107 Pedestrian view of Yanmar headquarters. Credits: Yanmar.

Figure 5.108 Detail of the Yanmar BIPV curtain wall. Credits: Yanmar.

Figure 5.109 Site plan of Yanmar headquarters. Source: Google Maps.

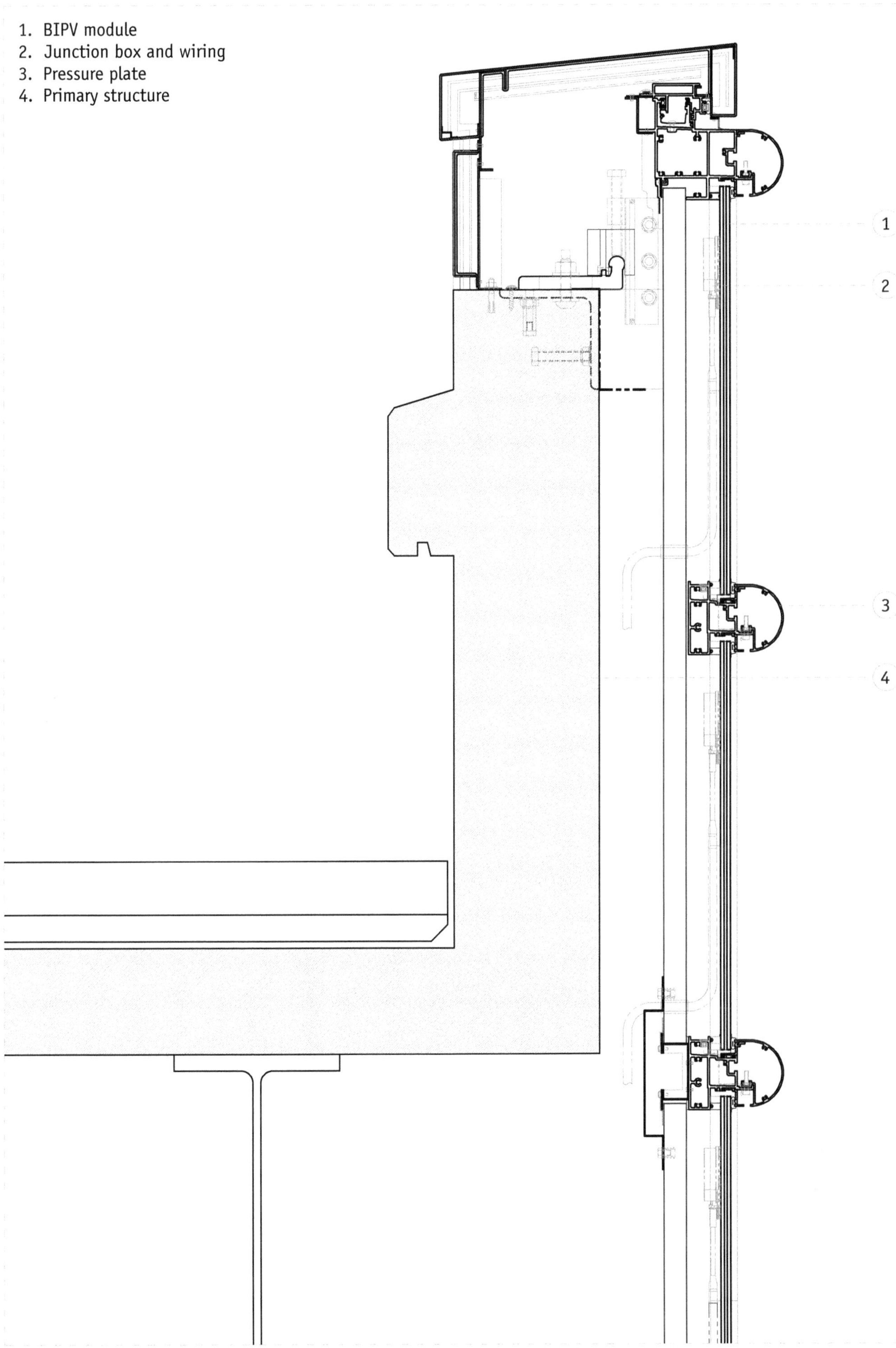

Scale 1:8

Figure 5.111 Yanmar headquarters. Cross-section drawing of the BIPV curtain wall. Credits: adapted by Ana Marcos Castro from Nikken Sekkei ©.

NCC headquarters

LOCATION	**Solna, Sweden (59° 22′ 0″ N)**
USE	**Office**
CLIENT	**NCC**
PV PRODUCER	**Cenit Vision by ISSOL**
ARCHITECT	**White Arkitekter**
ENGINEER	**NCC**
INSTALLER	**Fenestra Wieden**
YEAR OF CONSTRUCTION	**2023**

Project description: Located in the Municipality of Solna, just north of the City of Stockholm, the building was completed in 2019. It encloses 22,000 m² of office space and accommodates 800 workplaces. Several aspects of the design emphasise sustainability as core to this project, which has granted the building an "excellent" certification in the BREEAM rating system. With the health and wellness of workers in mind, while also considering circular engineering principles, responsibly sourced laminated veneer lumber (LVL) is the main material used for the structure and interior finishes. In the main atrium, daylight illuminates a sculptural staircase that welcomes visitors and enhances the movement and encounters of people, from the main lobby to the upper floors.

BIPV system key characteristics: The façade includes BIPV modules in dark grey and blue, providing a refined urban look to the southeast and southwest façades of the building. They are incorporated into the spandrel sections of the curtain wall. The colour is achieved through a semi-transparent film placed between two glass panes, making the solar cells invisible. The BIPV modules have been developed specifically for the project to satisfy design needs, as the product can be manufactured with varying size, shape, and thickness to conform with regulating norms or architectural requirements.

0.9 MWh/m²
Annual horizontal solar irradiation

Humid temperate – Cfb
Köppen climate classification

Curtain wall
BIPV application

New
Type of construction

Opaque coloured m-Si
PV technology

36.0 kW
BIPV array nominal power

60 kWh/m²
Annual electricity generation per BIPV module area

22.0 MWh
Annual electricity generation

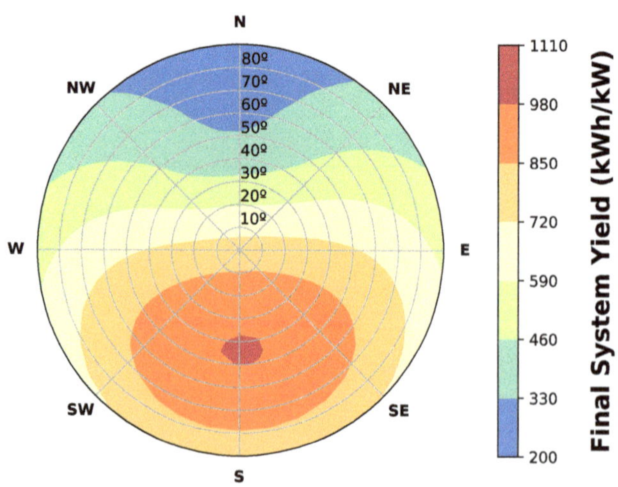

Figure 5.115 PV potential heatmap for Solna, Sweden. Credits: Ana Marcos Castro, CIEMAT.

Figure 5.112 Pedestrian view of NCC Headquarters. Credits: White Arkitekter AB.

Figure 5.113 Close-up of the NCC Headquarters BIPV curtain wall. Credits: White Arkitekter AB.

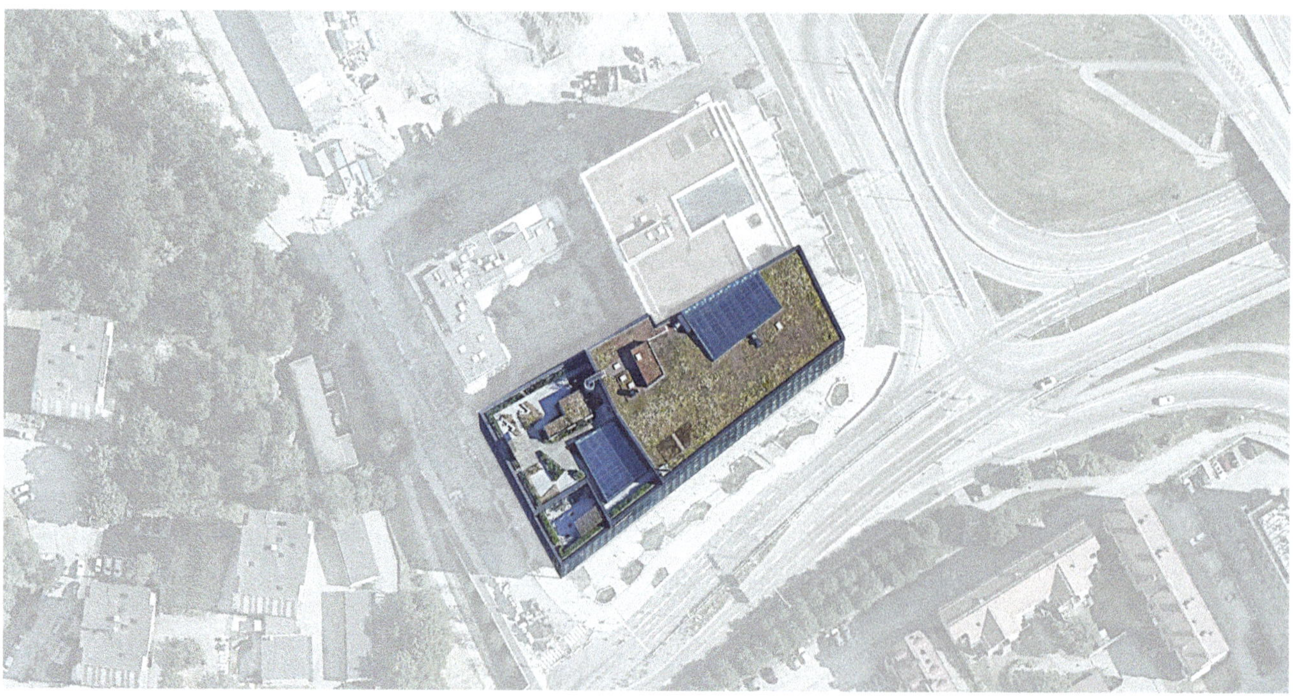

Figure 5.114 Site plan of NCC headquarters. Source: Google Maps.

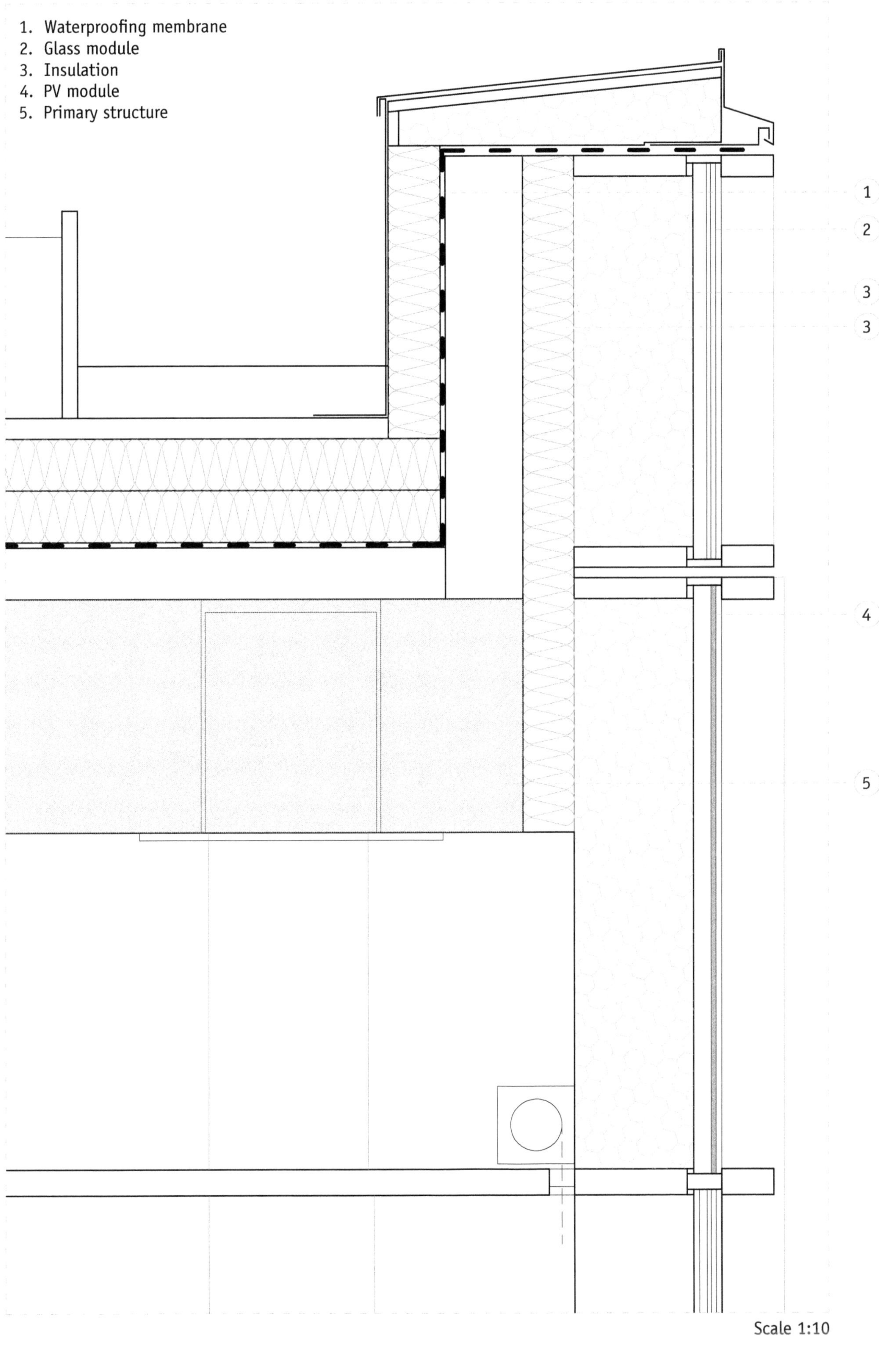

Figure 5.116 NCC headquarters. Cross-section drawing of the BIPV curtain wall parapet. Credits: adapted by Ana Marcos Castro from White Arkitekter AB ©.

Palazzo Lombardia

LOCATION	**Milan, Italy (45° 29′ 12″ N)**
USE	**Government & Retail**
PV PRODUCER	**EnergyGlass**
ARCHITECT	**Sistema Duemila Architettura e Ingegneria**
ENGINEER	**PEI Cobb Freed & Partners, Caputo Partnership Consortium**
INSTALLER	**ISA S.p.A.**
YEAR OF CONSTRUCTION	**2011**

Project description: Located in downtown Milan, Palazzo Lombardia is an 87,000 m² complex that houses 3,000 regional government employees and retail space. Inspired by the Lombardy mountains, the urban complex consists of four curvilinear buildings symbolising the mountainous skyline, and a 39-storey (161 m tall) skyscraper. The façade of the skyscraper is partially covered with BIPV that generates a portion of the building's electricity demand while a geo-exchange heat pump system is connected to the Martesana canal, meeting the demand for heating. In 2012, Palazzo Lombardia received the Best European Skyscraper of the Year award from the Council of Tall Buildings and Urban Habitat (CTBUH).

BIPV system key characteristics: The skyscraper has a 1,420 m² surface of semi-transparent BIPV curtain wall with a thermal transmittance of U = 1.3 W/(m²K) and a visible transmittance of TVIS = 34%. More than 200 unitised curtain wall units were assembled in a factory, shipped, and installed on-site, ensuring quality control and ease of installation. Each 3.5-m-tall curtain wall unit incorporates structural, hygrothermal, and electrical components in a single unitised modular solution. In addition to solar electricity generation, the BIPV curtain wall provides partial shading for the tower, reducing cooling demand during the cooling season.

1.1 MWh/m²
Annual horizontal solar irradiation

Humid temperate – Cfb
Köppen climate classification

Curtain Wall
BIPV application

New
Construction

Semi-transparent m-Si
PV technology

170.4 kW
BIPV array nominal power

73.5 kWh/m²
Annual electricity generation per BIPV module area

104.4 MWh
Annual electricity generation

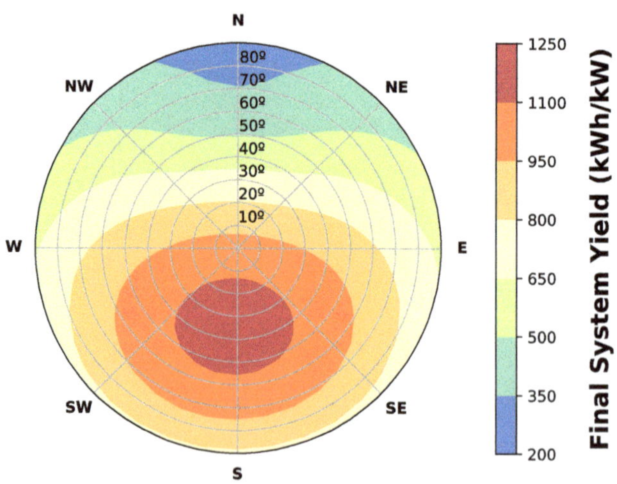

Figure 5.121 PV potential heatmap for Milan, Italy. Credits: Ana Marcos Castro, CIEMAT.

Figure 5.117 Eagle-eye view of the Palazzo Lombardia complex. Credits: Caputo & Partners.

Figure 5.118 BIPV unitised curtain wall façade at Palazzo Lombardia. Credits: DiARC.

Figure 5.119 Various stages of the BIPV unitised curtain wall assembly. Credits: ISA S.p.A.

Figure 5.120 Site plan of Palazzo Lombardia. Source: Google Maps.

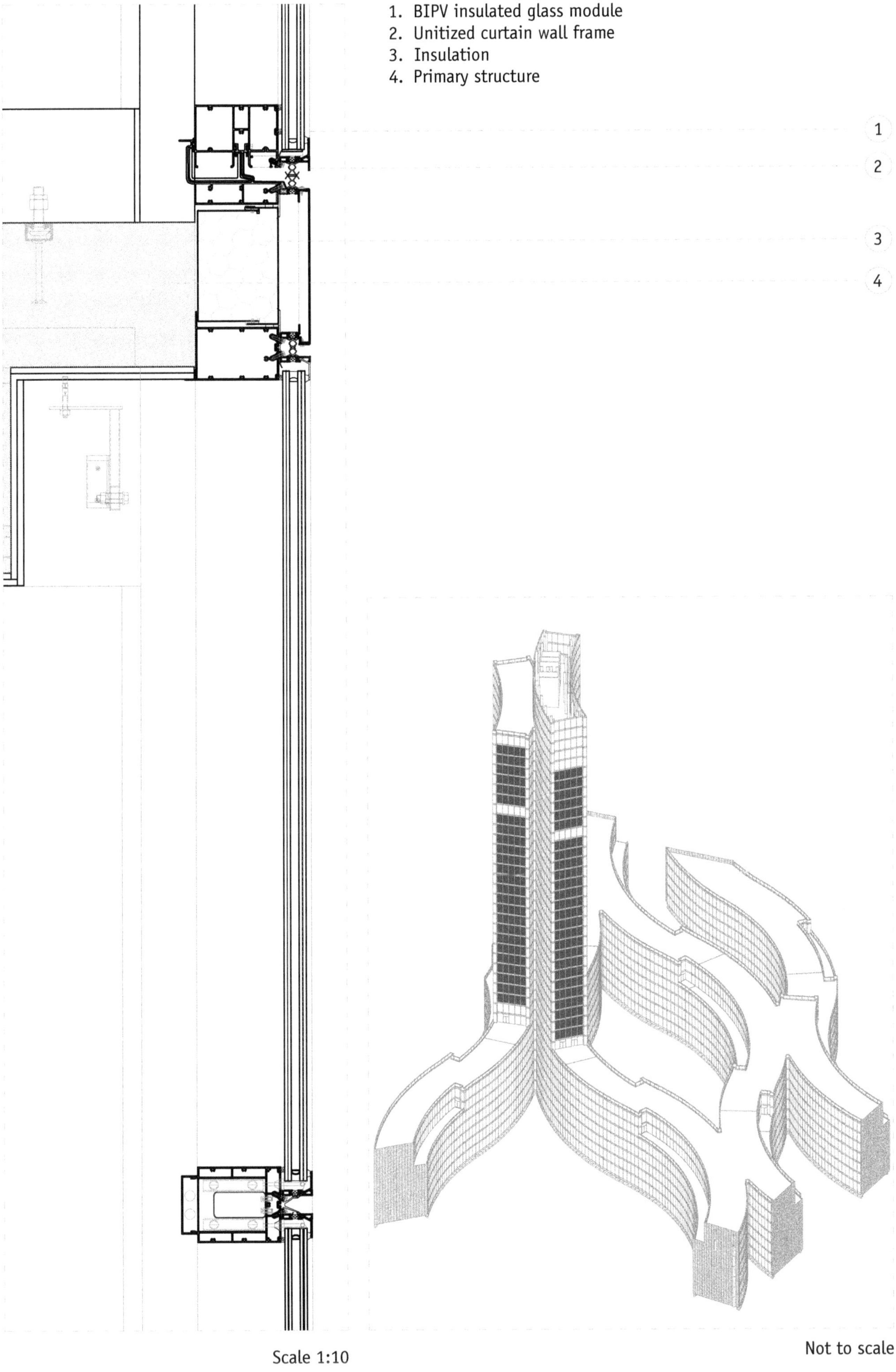

Figure 5.122 Palazzo Lombardia. Axonometric and cross-section drawings of the BIPV curtain wall. Credits: adapted by Ana Marcos Castro from Sistema Duemila Architettura e Ingegneria ©.

5.4 Externally integrated systems

5.4.1 Solar shading

General: When designed properly, exterior shading systems can significantly enhance the aesthetics of a building façade. The primary functions of a BIPV solar shading system comprise both solar electricity generation and sunlight control. This section looks at the design considerations of fixed BIPV louvers and overhangs ④.

Structural Design: The shading device and anchor points ② to the building must be designed to support the weight of the system and environmental loads such as wind, snow, and ice. Design must account for wind speeds specific to the building's location, and the shape, size, and location of the shading elements, including wind uplift. A spacing between the louvers and the external face of the façade will allow wind to partially flow through the installation and reduce wind pressure. Depending on the system design and structural requirements, the solar shading can be attached to the curtain wall mullions or cladding, anchored to the load-bearing building structure (e.g., concrete floor slab), or supported by an external superstructure ①.

Hygrothermal Performance: The shading systems should always be designed to deflect the elements such as rain and snow away from the building façade. A slope away from the façade is preferred while a slope towards it should be avoided. When the shading system is anchored to the load-bearing building structure, thermally broken brackets should be used to reduce thermal bridging.

Electrical Considerations: In the case of BIPV louvers, their spacing and depth should be carefully selected to prevent inter-louver shading at periods of high solar flux. When shading is unavoidable, the use of power optimizers or microinverters can help mitigate partial shading effects. The use of cable trays, clips, and ties serves to secure DC wiring ⑥ and minimise movement due to wind, which might otherwise lead to cable rattle and potential damage of wiring or connectors. For aesthetic purposes, the wiring can be hidden under cover caps.

Architectural Considerations: The latitude and orientation impact BIPV shading design and performance. Horizontal louvers and overhangs work best on near equatorial-facing façades (e.g., near south-facing in the northern hemisphere and near north-facing in the southern hemisphere) because of the higher solar angles in those directions. On the other hand, vertical BIPV louvers are well-suited for near east- and west-facing façades due to the lower angles of the rising and setting sun. While simple to implement, these basic guidelines are far from optimal. Thus, an annual solar shading analysis is recommended to assist designers in determining the following key design parameters: (1) tilt angle, (2) depth, and (3) spacing between louvers (if applicable). Most modern computer-aided design tools offer easy-to-use capabilities for this purpose. The parametric simulation study should aim to control solar gains and daylight, assess outdoor view, and maximise electricity generation. In the case of BIPV louvers, inter-louver shading should be avoided at times of high solar potential.

Sustainability: A BIPV shading device can be implemented in a new or an existing building with the intention to enhance solar gain and daylight control and maximise electricity generation. Designing a shading system with reuse in mind ensures a design that can adapt to evolving BIPV technologies or future aesthetic preferences.

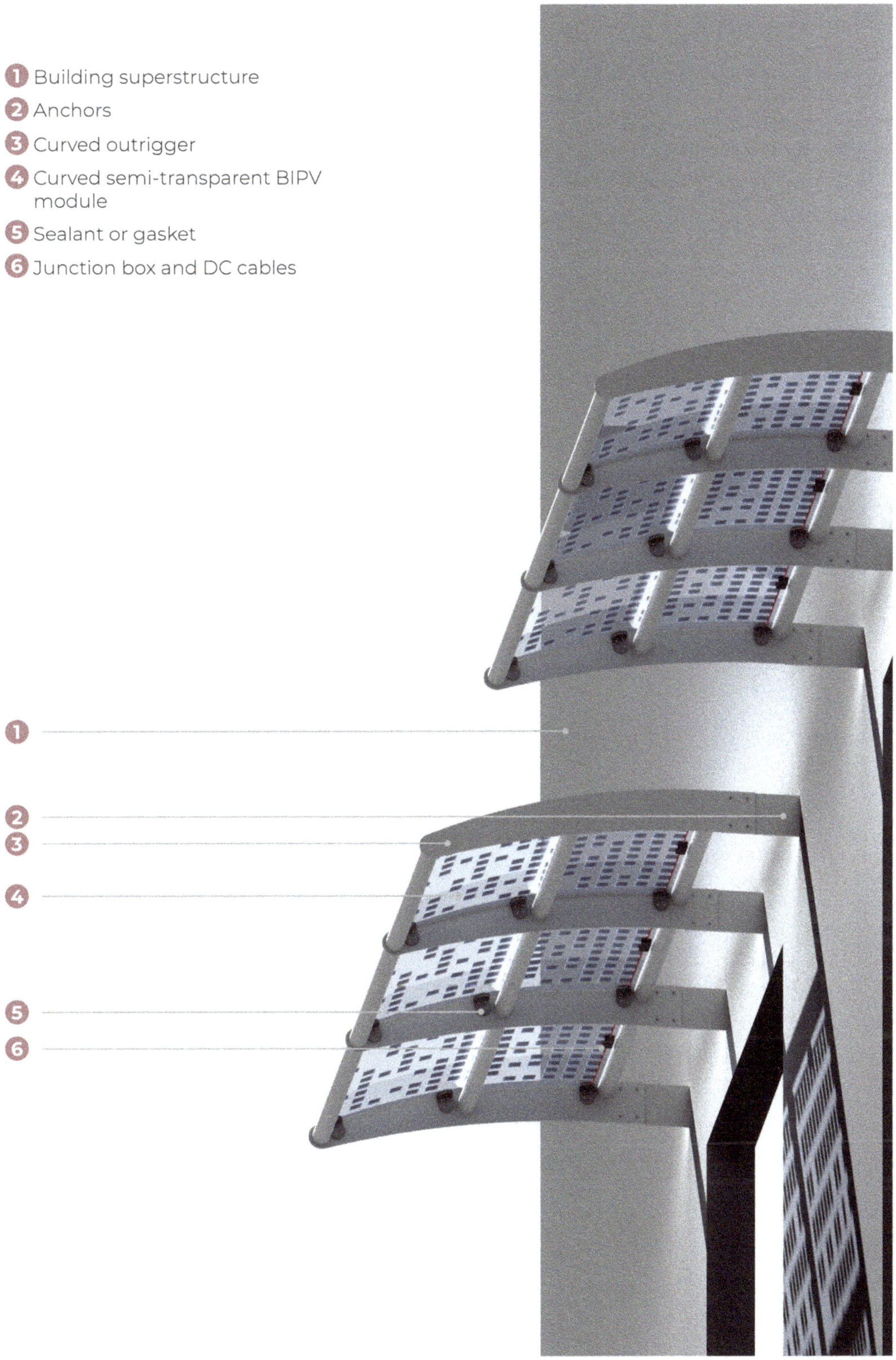

Figure 5.123 Axonometric representation of a horizontal BIPV overhang. Credits: Marley Dowling.

Gothenburg garage

LOCATION	**Gothenburg, Sweden (57° 39′ 41″ N)**
USE	**Garage**
PV PRODUCER	**Advanced Solar Power**
ARCHITECT	**Liljewall arkitekter**
ENGINEER	**Fasadsystem (Soltech Energy Sweden subsidiary)**
INSTALLER	**Fasadsystem**
YEAR OF CONSTRUCTION	**2021**

Project description: The project focuses on the construction of a solar façade on a new garage, owned by Wallenstam AB, providing charging points for electric vehicles in addition to generating electricity that is sold to the grid. An important aspect was the flow and renovation of air due to vehicle exhaust fumes. To tackle this issue, the project used an open façade design that enables natural airflow, which also benefits the photovoltaic modules by helping cool them down. Instead of designing a flat, continuous façade, the structure follows a 3D zig-zag design that creates gaps between each row of panels and allows constant airflow. In addition, the modules are semi-transparent to provide natural lighting to the inside of the garage. The colour pattern was requested to follow the work of Swedish textile designer Viola Gråsten, a pioneer in brightly coloured rya pile carpets and modern fabric designs. This inspiration translated into the colourful façade seen in the finished garage. The building is a statement that BIPV can be creative, innovative, and visually appealing – even for a less noble intended use building such as a garage.

BIPV system key characteristics: The base structure is a prefabricated 11-meter hot-dipped galvanised steel construction built by Soltech's subsidiary Fasadsystem that provides the desired angles for the PV modules. The modules are 1200 mm × 600 mm × 6.8 mm glazing units of semi-transparent thin-film cadmium telluride (CdTe) technology in four different colours: blue, green, orange, and terracotta red.

790 kWh/m²
Annual horizontal solar irradiation

Humid temperate – Cfb
Köppen climate classification

Solar shading
BIPV application

New
Type of construction

Semi-transparent coloured CdTe
PV technology

60 kW
BIPV array nominal power

21 kWh/m²
Annual electricity generation per BIPV module area

16.9 MWh
Annual electricity generation

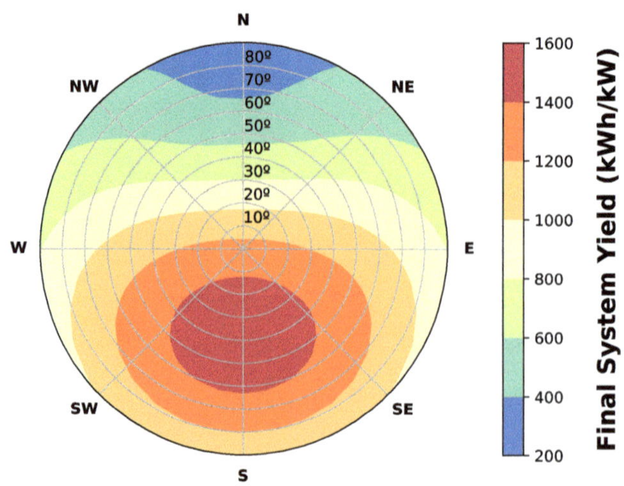

Figure 5.127 PV potential heatmap for Gothenburg, Sweden. Credits: Ana Marcos Castro, CIEMAT.

Figure 5.124 Coloured transparent BIPV shading at Gothenburg Garage. Credits: Soltech Energy Sweden.

Figure 5.125 Detail from the interior of Gothenburg Garage. Credits: Soltech Energy Sweden.

Figure 5.126 Site plan of the Gothenburg Garage. Source: Google Maps.

Not to scale

Figure 5.128 Gothenburg Garage. Axonometric drawing of the BIPV shading device. Credits: adapted by Ana Marcos Castro from Soltech Energy Sweden©.

Franklin University Student Residence

LOCATION	Sorengo, Switzerland (45° 59′ 52″ N)
USE	Residence
PV PRODUCER	SUNAGE
ARCHITECT	Flaviano Capriotti Architetti
ENGINEER	Aziende Industriali di Lugano (AIL)
INSTALLER	Kummler + Matter, Poretti Gaggini Consortium
YEAR OF CONSTRUCTION	2023

Project description: The new Students' Residence at Franklin University consists of two building volumes with different functions: the dynamic and bright volume of educational and communal spaces using U-profile glass with a satin finish on its façade, and the massive and introverted residential one using pigmented concrete. Dynamic BIPV vertical louvers are controlled to optimise solar gains and occupant comfort year-around. The façade is one of the first dynamic BIPV projects in Europe, designed to meet sustainability, thermal, and visual comfort requirements.

BIPV system key characteristics: The dynamic BIPV louvers have been specifically engineered and manufactured for this project; they consist of laminated tempered glass with a white satin finish (*bianco traffico* by Suncol company). The louvers are fastened on a metal substructure and controlled by one motor for every three blades: an algorithm tracks the sun on a single vertical axis and optimizes the orientation of the BIPV louvers to maximise electricity generation and comfort. The solution reduces self-shading mismatch by subdividing the internal electric circuit of each BIPV louver into two "electrically independent" columns. This design attribute increases the annual electricity yield of the BIPV system by 20% compared to a single-circuit BIPV louver configuration.

1.3 MWh/m²
Annual horizontal solar irradiation

Continental subarctic – Dfc
Köppen climate classification

Solar shading
BIPV application

New
Type of construction

Coloured opaque mc-Si
PV technology

18 kW
BIPV array nominal power

102 kWh/m²
Annual electricity generation per BIPV module area

18.7 MWh
Annual electricity generation

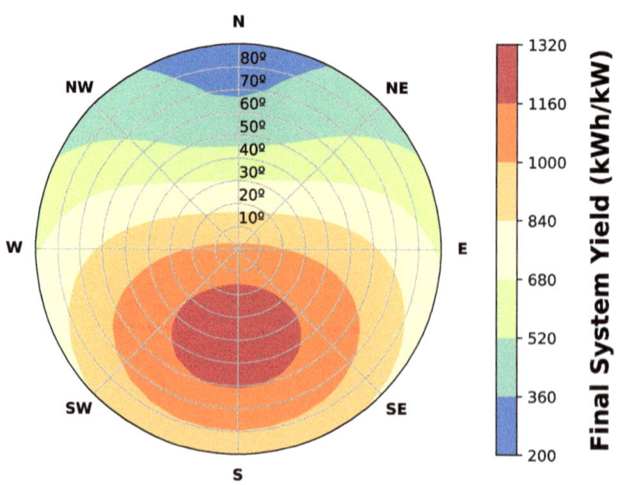

Figure 5.132 PV potential heatmap for Sorengo, Switzerland. Credits: Ana Marcos Castro, CIEMAT.

Figure 5.129 Pedestrian view of Franklin University Student Residence. Credits: Franklin University.

Figure 5.130 Installation of the white movable BIPV solar shading louvres. Credits: Franklin University.

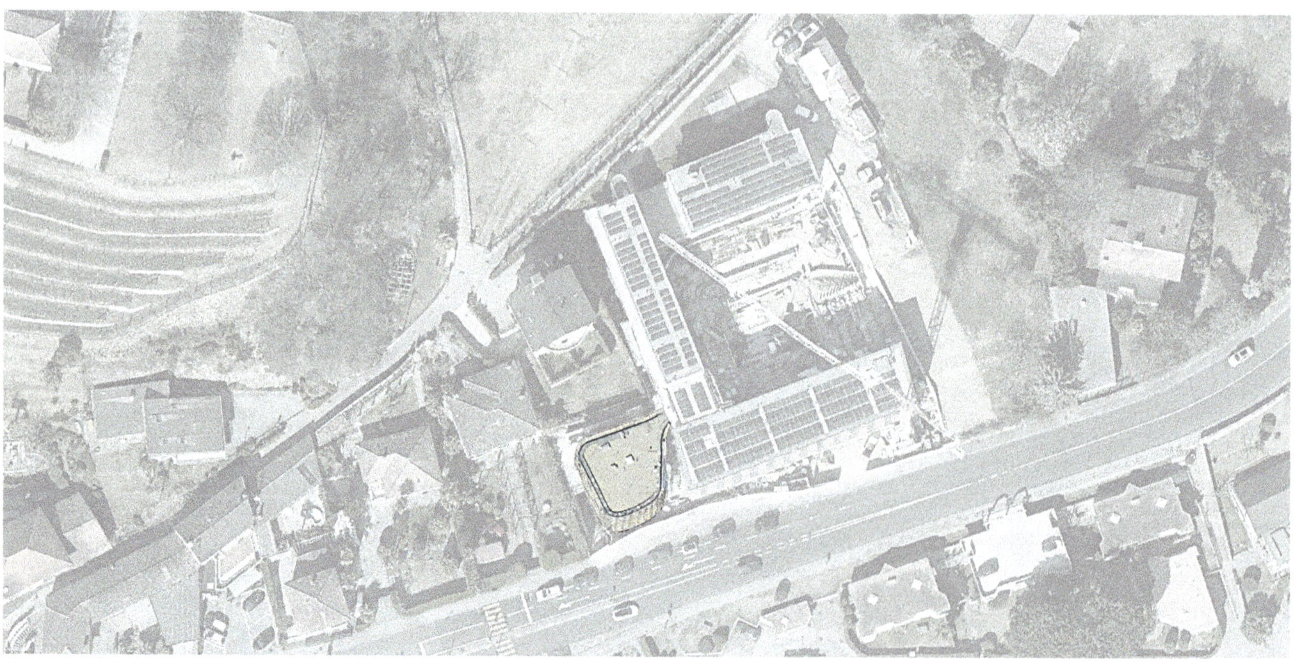

Figure 5.131 Site plan of Franklin University Student Residence. Source: Google Maps.

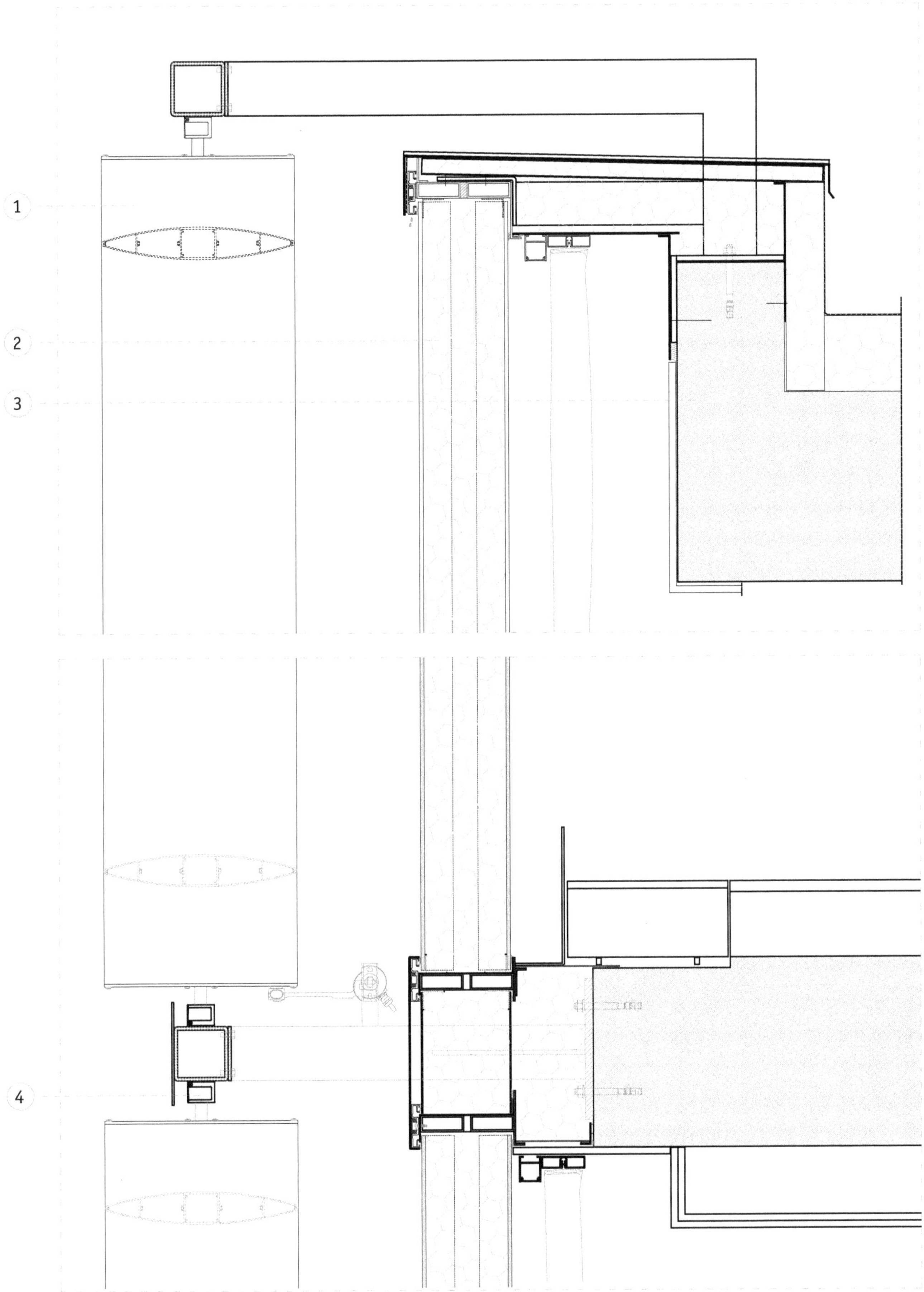

Scale 1:10

1. BIPV sunshade louver
2. Insulation
3. Primary structure
4. Motor

Figure 5.133 Franklin University Student Residence. Cross-section drawing of the BIPV solar shading system. Credits: adapted by Ana Marcos Castro from Thema ©.

5.5 Infrastructure

5.5.1 Parking Canopy and Transit Shelter

General: PV canopies and transit shelters are infrastructure designed to provide shade and shelter from the elements, while they can be used to power system services (e.g., lighting, information displays, Wi-Fi), nearby buildings, and electric vehicle charging stations. Typically installed in parking lots, recreational spaces and parks, bus and railway stations, and ferry terminals, solar canopies and shelters make efficient use of otherwise unused or underutilised spaces and can serve as public educational tools, raising awareness about solar energy use and sustainability in the built environment.

Structural Design: Safety is paramount, so the structure must adhere to local codes and regulations. The superstructure ① should be oversized to support the weight of the PV modules ③, withstand environmental loads (seismic, wind, rain, ice and snow), and resist unexpected live and impact loads (e.g., people hanging from or climbing on top of a solar shelter, vehicle impact). The construction materials should be durable and corrosion-resistant to ensure the structure's long-term integrity. Typically, these structures incorporate appropriate foundations, such as concrete footings or piles, to anchor the canopy securely to the ground.

Hygrothermal Performance: Effective rain, snow, and ice management is crucial for extending the canopy or shelter's lifespan, enhancing user comfort, and ensuring the infrastructure's efficient and safe operation. To achieve this, it is essential to employ sloping PV roofs and implement a gutter system ⑦ that directs rainwater away from the foundations. In the case of a butterfly roof (ideal for near east-/west-oriented installations), it is advised to use a continuous valley gutter along the canopy's centerline. In snowy regions, butterfly roof canopies tend to retain snow and ice for extended periods, reducing PV roof electricity yield in winter. To mitigate this, a single-pitch solar roof canopy and shelter (ideal for near south-oriented installations) design with gutters located at the roof edge can be employed. These gutters collect rainwater and channel it to a downpipe and divert it away from the structure. See BIPV roofs and skylights section for more information on how to achieve watertight connections between PV modules.

Fire Safety: The use of non-combustible materials can assist with resistance to fire and structural integrity. Depending on the design, the use of an automatic sprinkler protection system and a rapid shutdown switch may be required by the local fire code and regulations. For more information, see section 6.3, Safety Aspects.

Electrical Considerations: The junction boxes and DC wiring ⑥ should be concealed behind cover caps, making them inaccessible to the public but still reachable by the maintenance crew. An outdoor-rated electrical enclosure houses the inverters. For locations and climates with significant ground albedo (e.g., snow ground coverage during winter months or proximity to a body of water that can act as a "mirror," reflecting the sunlight toward the rear side of the canopy), the use of bifacial PV modules is recommended to increase the annual electricity yield by 10% to 25% [11]. When electric vehicle (EV) charging stations are incorporated, the PV carport is typically connected to the power grid, rather than directly to the charging stations. The PV canopy generates solar electricity and feeds it into the grid. The EV charging stations then draw power from the power grid as needed to charge the electric vehicles. This setup allows for flexibility in distributing the generated solar electricity to various loads, including the charging stations, while ensuring a continuous charging capacity even when solar generation is insufficient, such as during cloudy days or at night.

Architectural Considerations: The use of semi-transparent PV modules ③ on the canopy or shelter can enhance daylight utilisation, reduce the need for electric lighting during the day, and create a more pleasant luminous environment for the users. When combined with an aesthetically appealing exposed structure, a solar canopy or a shelter can serve as a focal point within a city. Such solar structures often capture the attention of users and visitors alike, becoming iconic landmarks and contributing to the city's identity. For further inspiration, see the Arrival Center in Schönbrunn Palace, in Vienna.

Sustainability: PV canopies can contribute on the reduction of carbon emissions in the built environment, support the electrification of the transportation sector, and most importantly, increase public awareness of renewable energy and sustainability.

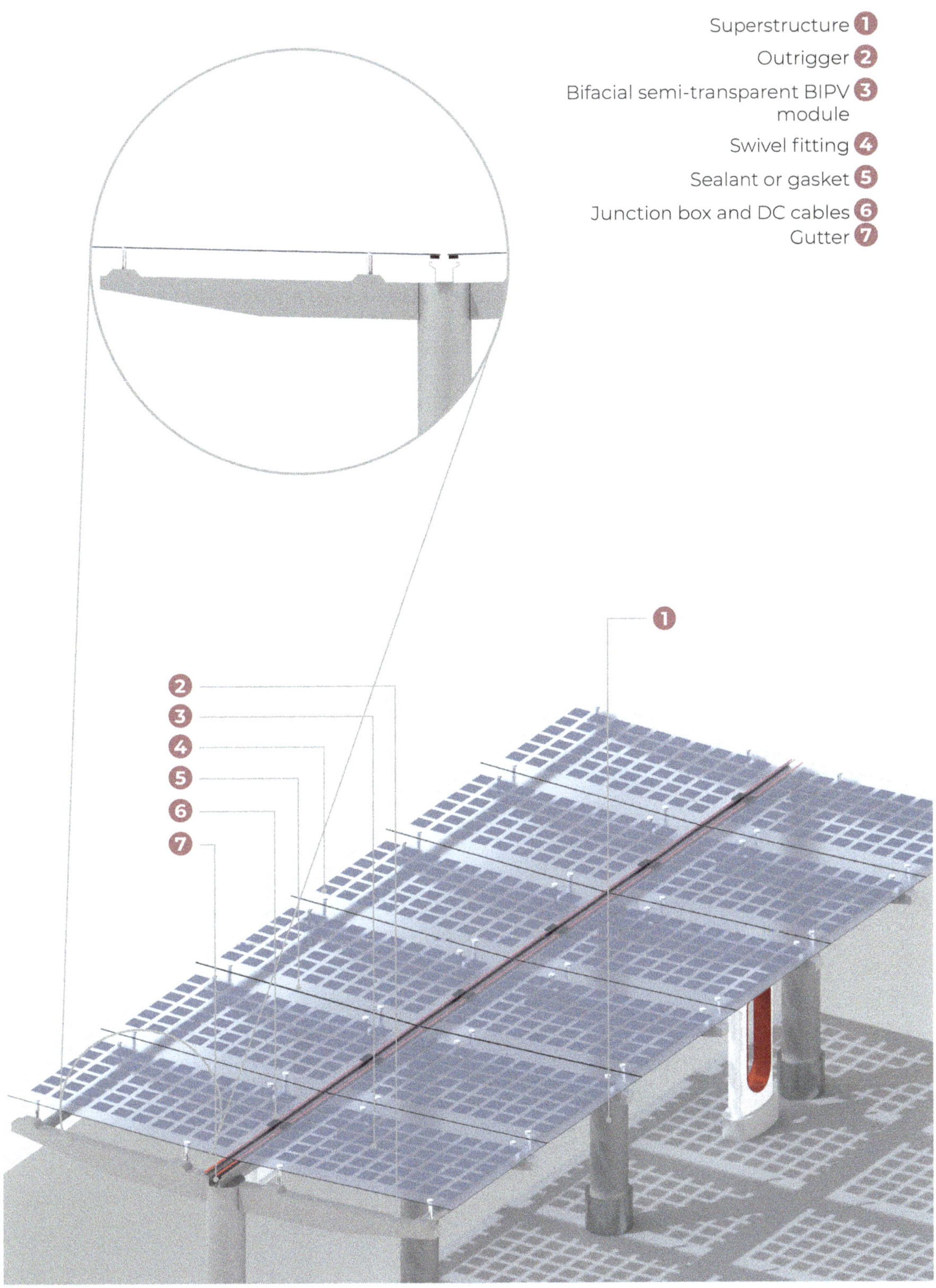

Figure 5.134 Axonometric representation of a BIPV canopy. Credits: Marley Dowling.

Arrival Center in Schönbrunn

LOCATION	Vienna, Austria (48° 11′ 10″ N)
USE	Carport
PV PRODUCER	Ertex Solar
ARCHITECT	FCP Fritsch, Chiari & Partner ZT GmbH
ENGINEER	HESS Stahlbau & Montage GmbH
INSTALLER	Ertex Solar
YEAR OF CONSTRUCTION	2020

Project description: Schönbrunn Palace in Vienna is known and appreciated all over the world, which explains the influx of tourists throughout the year. With the construction of two new facilities, Schönbrunn Palace now has its own reception area, the Arrival Center, to provide better and more convenient access for national and international visitors.

The first phase of construction included the reception centre, bus parking, a covered exit, and green areas, while the second phase of construction, completed on June 29, 2020, provided visitors with 230 parking spaces. Visitor comfort, satisfaction, and easy access were the primary objectives for this solar project, ensuring a relaxed arrival for visitors driving to this world-renowned landmark.

BIPV system key characteristics: The PV system has a total surface of 576 m². It's located in the northwestern part of the car park and was designed as a carport-like construction. A total of 120 laminated safety glass modules with an impressive size of 960 × 5,000 mm are integrated into its roof. Every single PV module has a power of 600 W each. The semi-transparent PV installation ensures that visitors are protected from the weather elements during rainy or snowy days while providing partial shading during sunny days. Furthermore, the plant supplies electricity for 10 integrated EV charging stations.

1.1 MWh/m²
Annual horizontal solar irradiation

Humid temperate – Cfb
Köppen climate classification

Canopy
BIPV Application

New
Type of construction

Semi-transparent m-Si
PV technology

72 kW
BIPV array nominal power

117.4 kWh/m²
Annual electricity generation per BIPV module area

67.6 MWh
Annual electricity generation

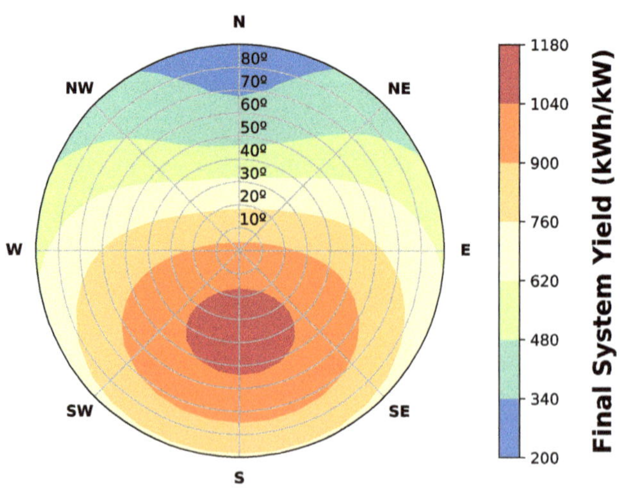

Figure 5.139 Solar potential heatmap for Vienna, Austria. Credits: Ana Marcos Castro, CIEMAT.

Figure 5.135 BIPV car canopy at the arrival centre in Schönbrunn. Credits: Francesco Frontini, SUPSI.

Figure 5.136 Close-up view of the BIPV car canopy at the arrival centre in Schönbrunn. Credits: Francesco Frontini, SUPSI.

Figure 5.137 Night view of the BIPV car canopy at Schönbrunn. Credits: Ertex solar.

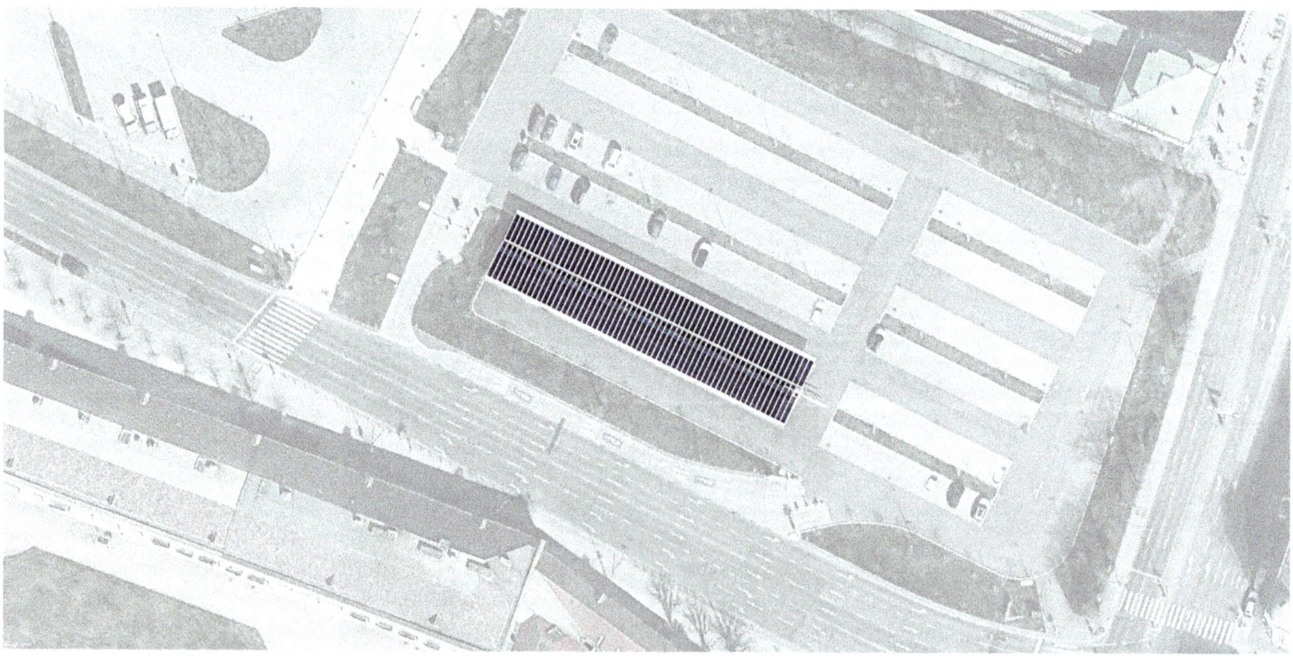

Figure 5.138 Site plan for the arrival centre in Schönbrunn. Source: Google Maps.

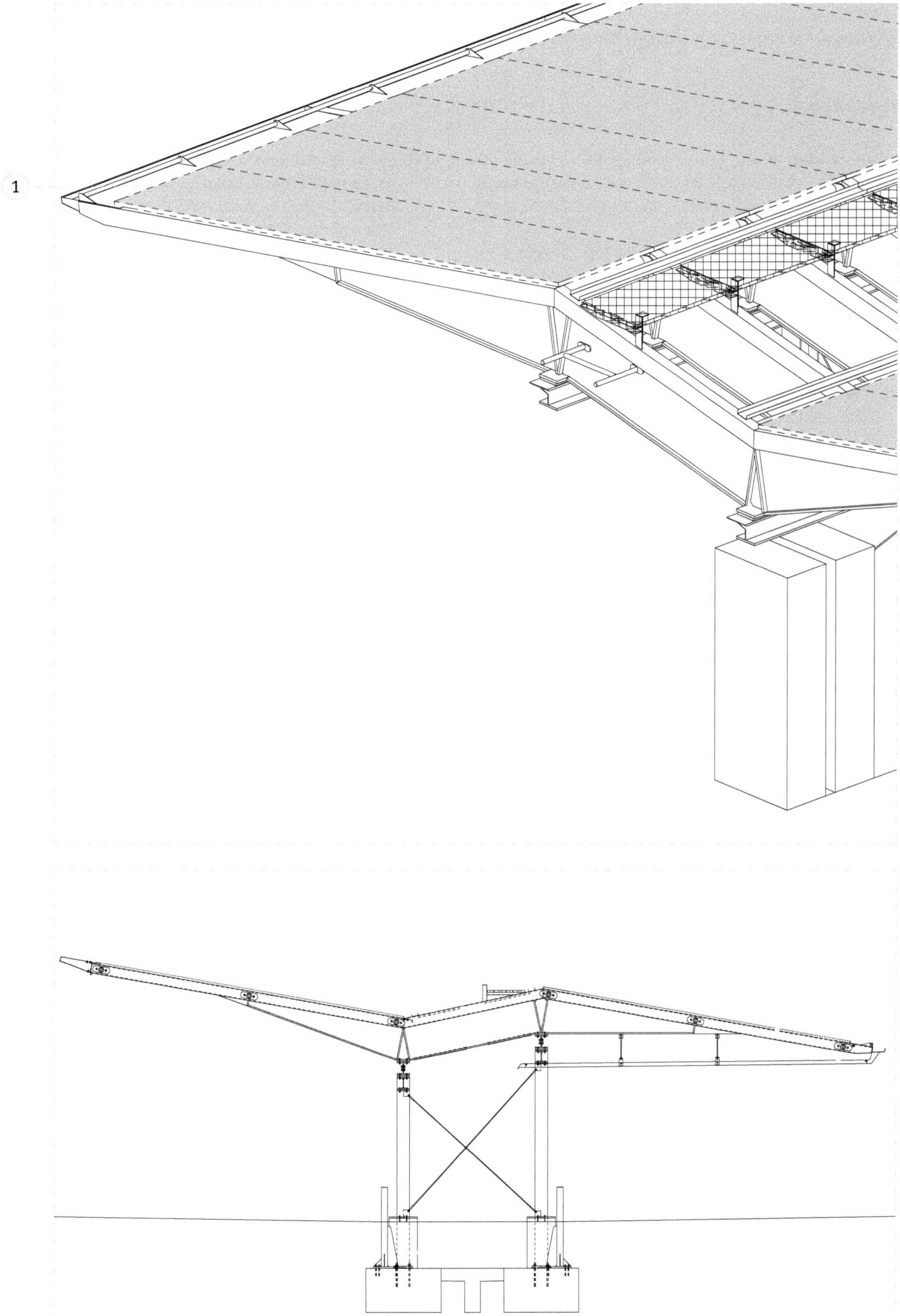

Figure 5.140 Axonometric and cross-section drawings of the BIPV car canopy at the arrival centre in Schönbrunn. Credits: adapted by Ana Marcos Castro from HESS Stahlbau & Montage GmbH ©.

Not to scale

References

[1] P. Bonomo, et al., "Categorization of BIPV applications," *IEA-PVPS T15–12:2021*. IEA PVPS, 2021, https://iea-pvps.org/wp-content/uploads/2021/09/IEA-PVPS-T15-12_2021_BIPV-categorization_report.pdf.

[2] T. Reijenga, et al., *Successful Building Integration of Photovoltaics: A Collection of International Projects*. International Energy Agency, 2020.

[3] I. B., *Gebäudeintegrierte Photovoltaik: Architektonische Integration der Photovoltaik in die Gebäudehülle*. Rudolf Müller, 2002.

[4] PACTE, "Systèmes photovoltaïques par modules rigides intégrés en couverture (Report DGALN/DHUP – 978-2-11-151809-4)," *Le Programme d'Action pour la qualité de la Construction et la Transition Energétique*. PACTE, 2017, https://www.proreno.fr/documents/systemes-photovoltaiques-par-modules-rigides-integres-en-couverture.

[5] CanmetENERGY, *Photovoltaic Ready Guidelines, Version 2.0. (Report M154–122/2019E-PDF)*. Natural Resources Canada, 2019.

[6] R. A. Agathokleous and S. A. Kalogirou, "Double skin façades (DSF) and building integrated photovoltaics (BIPV): A review of configurations and heat transfer characteristics," *Renew. Energy*, vol. 89, pp. 743–756, 2016.

[7] J. Carmody, et al., *Window System for High-Performance Buildings*. Norton Professional Books, 2004.

[8] B. Weller, et al., *Photovoltaics: Technology Architecture Installation*. Detail Practice, 2010.

[9] N. Martín Chivelet and I. Fernández Solla, *La envolvente fotovoltaica en la arquitectura: criterios de diseño y aplicaciones*. Reverté, 2007. ISBN: 978-84-291-2112-4.

[10] *ASHRAE Standard 90.1–2022 – Energy Standard for Sites and Buildings Except Low-Rise Residential Buildings*. ASHRAE.

[11] W. Gu, et al., "A comprehensive review and outlook of bifacial photovoltaic (bPV) technology," *Energy Convers. Manag.*, vol. 223, p. 113283, 2020.

CHAPTER 6

Operation and maintenance of BIPV systems

Nuria Martín Chivelet, Alexander Astigarraga, Maximilian Lugmair, and Simon Boddaert

6.1 Introduction

Operation and maintenance (O&M) of a BIPV system encompasses the tasks required to keep the system functioning optimally and extend its expected lifespan. Incorporating O&M during the design, engineering, and construction phases of a BIPV system is essential. It enhances the system life cycle management, minimizes potential risks, and reduces life cycle costs.

There are established O&M guidelines and standards for conventional photovoltaic systems [1–3] summarised in Table 6.1, but BIPV systems lack comprehensive literature and adapted standards. This chapter aims to address this void by highlighting key considerations for BIPV systems.

6.2 Main components of an O&M plan

To facilitate the O&M of a BIPV system and minimise unexpected future costs, it is recommended to choose system components that come with comprehensive manufacturer warranties, offer low failure rates, and are designed for easy replacement. A comprehensive O&M plan should be prepared, detailing the information listed as follows. This O&M plan should be readily available to the owner or upon request, and a copy should be kept in a specified location (e.g., electrical room):

Table 6.1 IEC technical specifications and standards providing O&M rules and guidelines for photovoltaic systems, not specifically for BIPV.

Photovoltaic (PV) systems – Requirements for testing, documentation and maintenance	IEC 62446-1	Part 1: Grid connected systems – Documentation, commissioning tests and inspection
	IEC 62446-2	Part 2: Grid-connected systems -Maintenance of PV systems
	IEC TS 62446-3	Part 3: Photovoltaic modules and plants – Outdoor infrared thermography
	IEC TS 62446-4	Part 4: Electroluminescence measurement of photovoltaic arrays
Terrestrial photovoltaic (PV) systems – Guidelines for effective quality assurance in PV systems installation, operation and maintenance	IEC TS 63049	
Photovoltaic power systems – Reliability practices for operation	IEC TS 63265	
Photovoltaic (PV) array – On-site measurement of current-voltage characteristics	IEC 61829	

Contact Information

Contact information of the responsible party
Contact information of access and maintenance personnel

DOI: 10.1201/9781003432241-6

System Description

System documentation. For the electrical part, IEC 62446 applies.
Electrical drawings of the BIPV system(s) as installed, including single line diagram(s)
BIPV location within the building envelope and its category (e.g., BIPV roof, façade)
Description of surrounding conditions (i.e., potential shading and human activity both within and around the building)
System description and list of all equipment such as manufacturer, model, serial number, replacement supplier details, and where each item is situated in the building, including rapid shutdown switches. Inventory of on-site readily available spare parts.
Maximum power of the PV generator, associated with specific accreditation low/high voltage DC and AC
A sizing study, which includes power predictions and the assumptions behind these calculations

O&M Description

Description of the monitoring system in place and data collection methodology (e.g., network-connected inverters for remote monitoring and reset)
BIPV system performance indicators used to assess the system
Alert protocols based on monitoring feedback, error notifications, or owner complaints
Repair or replacement criteria for each component and post-repair testing procedures
Budget for the O&M plan, including costs for monitoring and diagnostics, preventive and corrective maintenance, and contingency provisions for high-cost replacements or repairs
Plan for periodic check and replacement of energy storage batteries (if applicable)

Components Documentation

Manufacturers' warranties, including power generation guarantees, their duration, and suggested preventive measures (e.g., cleaning BIPV module surfaces)
Flash test reports showing the current-voltage characteristics of each BIPV module under standard test conditions, issued by an accredited laboratory. Additionally, having an electroluminescence image for each module is advantageous.
Valid safety compliance certificates
This list ensures comprehensive O&M coverage, minimising potential issues and facilitating reliable operation. Note that the *IEC 62446: Photovoltaic (PV) systems – Requirements for testing, documentation and maintenance* standard is also relevant to BIPV system commissioning and O&M. The standard outlines the necessary documentation, required commissioning tests, and inspection criteria for grid-connected systems that can be useful for future inspections, maintenance, and modifications.

6.3 Safety aspects

From the electrical point of view, safety requirements and recommendations for PV systems and modules apply to BIPV systems. However, the *IEC 63092: Photovoltaics in buildings* standard suggests additional safety measures specific to BIPV. Another crucial standard for PV modules is the *IEC 61730: Photovoltaic (PV) module safety qualification*. This standard covers construction and testing requirements for PV modules and is equally applicable to BIPV modules.

Applicable to BIPV systems is also the IEC 60364: *Low-voltage electrical installations* standard, with several related parts: part 1: fundamental principles, assessment of general characteristics, definitions; part 4: protection for safety; part 5: selection and erection of electrical equipment; part 6: verification; part 7: requirements for special installations or locations (e.g., section 712: solar photovoltaic power supply systems, section 722: supplies for electric vehicles), and part 8: functional aspects.

Typically, national codes and guidelines attain the safety requirements of *IEC 61730* and *IEC 60364* for modules and low-voltage installations. The European Union has adopted both standards as Harmonised Documents, which means that they are enforceable. According to the United States National Electrical Code (NEC), PV and BIPV system circuits installed on or in buildings shall include a rapid shutdown function to reduce shock hazard for emergency responders. Rapid shutdown gives the firefighters a way to disconnect the DC power of the PV generator to the power conditioning equipment (inverter). This good practice is also addressed under the international standard *IEC 63257*.

Beyond electrical safety, local codes, guidelines, and a range of ISO standards address safety related to structural design and performance evaluation of structural and non-structural components, including classification and testing of safety glass in buildings that apply to BIPV modules.

Only qualified personnel should handle BIPV installations to ensure safety and quality standards. The installation personnel should be certified as PV installers and hold qualifications specific to BIPV applications. Proper insurance and certificates are mandatory. Additionally, maintenance companies must ensure their workers' safety against potential hazards and accidents. Therefore, personal protective equipment should include protection against risks associated with the construction, commissioning, and operation of the BIPV systems. Safety precautions should include guardrails or other edge protections, maintenance pathways, safety nets, temporary platforms, marked hazard zones, and proper use of harness systems, lifelines, and ladders when working at elevated heights.

6.4 BIPV system operation

The O&M service provider is responsible for operating the BIPV plant following the respective national grid code. The requirements provided by the grid operator are usually: power quality, voltage regulation, and management of active and reactive power. The specificities and quality requirements depend on the voltage level of the grid. The O&M service provider is responsible for monitoring quality.

6.4.1 Operation checklist

The following checklist serves as a guide for operating BIPV systems. It is recommended to consider specific requirements and recommendations from the manufacturer or industry experts and undergo relevant training to ensure optimal system performance and safety. There should be a logbook that records the history of intervention. The following preliminary questions can help identify if the operation is being correctly assessed:

- Is the BIPV system performance actively monitored and recorded?
- Is the performance trend analysis done continuously or at regular intervals?
- Does the monitoring provider offer forecasting? If so, what are the forecast horizon, time detail, update rate, and precision?
- Are there specified response times for corrective maintenance in the contract?
- Is the BIPV module degradation rate specified?
- Are all service contracts and insurance policies kept updated?
- Are all required operating permits, compliance documents, and licenses kept up to date?

For optimal BIPV operation, adhere to the following guidelines that address inspection and maintenance, energy generation monitoring, wiring and connection inspection, inverter verification, documentation, training and education, shading analysis, safety, regular overall performance review, and compliance with local codes, regulations, and standards.

6.4.1.1 BIPV modules inspection and maintenance

Module surfaces, connections, and seals should be routinely inspected. As necessary, dust and debris should be cleaned using recommended products.

6.4.1.2 Energy generation monitoring

Energy output should be tracked using a monitoring system, highlighting potential issues. Actual system performance should be regularly compared with projections. If possible, a fault detection system should be utilised through the Energy Management System (EMS).

6.4.1.3 Wiring and connection check

BIPV wiring and connections should be checked for wear or damage. Firm and secure connections should be ensured for optimal electricity flow.

6.4.1.4 Inverter verification

Inverter performance and operational status should be monitored. Any alerts or anomalies should be promptly reviewed and acted upon.

6.4.1.5 Shading analysis

Periodic assessments should be conducted for obstructions blocking sunlight on the BIPV modules. Shading issues, such as overgrown vegetation, should be removed.

6.4.1.6 Documentation

Thorough records, including inspection reports and maintenance activities, should be documented. All system repairs or upgrades should be logged.

6.4.1.7 Training and education

Staff should be trained and kept up-to-date on BIPV system operations and maintenance. Personnel should be updated with the latest technologies and procedures.

6.4.1.8 Safety

Safety protocols should be followed to during maintenance and inspections. All electrical components should meet safety standards.

6.4.1.9 Regular performance review

The overall efficiency and performance of the BIPV system should be periodically evaluated. Areas for improvement or optimisation should be identified.

6.4.1.10 Regulatory compliance

The system should comply with all relevant codes, standards, and regulations. The system should be adjusted as necessary to meet any regulatory changes.

6.4.2 Data monitoring

Nowadays, energy production data are readily available and typically free of charge on the portals provided by inverter manufacturers. These portals offer access to a substantial portion of the inverter's variables, which can be easily checked and downloaded. This information serves both battery management system (BMS) and O&M purposes, including building-grid interaction, inverter status, battery status, module-level performance data (when optimizers or microinverters are installed), and more. Generally, these data are sufficient for basic system control and monitoring. However, the data frequency from the inverter's portal can be a limitation, ranging from 5 to 15 minutes.

Slow data frequency may hinder the detection or visualisation of shadow evolution on the PV modules' surface, a common issue in BIPV systems. A solution is a parallel monitoring system that enables the data collection at a higher frequency communicated to the inverter. If not, a last option could be directly measuring variables such as current and voltage to help detect shading or faults.

Main parameters to measure are the plane-of-array irradiance (G_{POA}), PV module temperature, AC output voltage, current, and power and the output power factor (requested by the utility). Also recommended is to measure DC output voltage, current, and power coming from the array, global horizontal irradiance, ambient temperature, and wind speed. When selecting an inverter, it is recommended to check if it includes the option of measuring irradiance and module temperature data. Unfortunately, these features are not yet widespread, so conducting market research may be necessary. In any case, a parallel monitoring system can measure these parameters and more, if required.

Aesthetics plays an important role in BIPV projects, so care should be taken when installing monitoring sensors and their cabling, trying to make them not visible. While fully instrumenting a BIPV façade may not be cost-effective, there should be a sufficient number of sensors strategically located to cover the monitoring needs. Although IEC 61724-1 can guide the placement of pyranometers, the standard does not specifically address BIPV systems, where irradiance at the plane-of-array tends to be non-uniform due to varying planes, heights, shading from the surroundings, etc. Pyranometers and temperature sensors are available in various sizes, technologies, and price ranges, offering flexibility for integration into the system (Figure 6.1), although with different technical characteristics and limitations.

Figure 6.1 Two examples of solar radiation sensors location: on the plane of a BIPV façade (left) and integrated into a BIPV module, developed under the BIPV Boost project (right).

Photos: courtesy of H.R. Wilson, Fraunhofer ISE (left) and A. Astigarraga, Eurac Research (right).

6.4.3 Energy performance metrics

Performance metrics for a PV system are listed in standard IEC 61724-1. According to this document, the most appropriate metric depends on system design, user requirements, and contractual obligations. Several performance metrics need to first measure (calculate) the energy output of the BIPV system, the most common being:

Final System Yield (Y_f): ratio of the net AC energy output (E_{PV}) of the BIPV system over the rated PV array DC power (nominal PV array power, P_0). It can refer to any reporting period, typically, a month or a year. Usually, it is expressed in kWh/kW:

$$Y_f = E_{PV}/P_0 \tag{6.1}$$

Performance Ratio (*PR*) is the ratio of the actual over the theoretically possible energy yield of an ideal installation, and it is calculated using equation 2.3. It indicates the overall effect of losses on the system output power due to both module temperature and system component inefficiencies or failures.

Note that the rating-based performance metrics are relatively simple to calculate but may omit known factors that cause system power output to deviate from expectations based just on the nominal rating alone. Performance index based on a detailed system model is thus more convenient to explain those deviations.

IEC TS 61724-2 outlines a methodology for measuring and assessing the power generation of a particular PV system with the aim of evaluating the performance quality of the PV system. This test is designed to be conducted over a relatively brief duration, typically a few days with ample sunshine. This standard procedure allows for a comparison between the actual measured power output and the anticipated power output of a PV system on days characterised by significant sunlight.

IEC TS 61724-3 establishes a methodology for measuring and evaluating the energy generation of a particular PV system in relation to the anticipated electrical energy production based on the prevailing weather conditions as defined by the parties involved in the test. The energy production is specifically assessed during operational periods, while periods of system inactivity are quantified as part of an availability metric. This is a standardised procedure for comparing the actual measured electrical energy output with the expected electrical energy output of the PV system.

Finally, the self-consumption index (*SCI*) and the self-sufficiency index (*SSI*), discussed in detail under section 2.5, are useful parameters to evaluate the evolution of the interaction between BIPV system generation and building consumption from the grid.

6.5 BIPV system maintenance

6.5.1 BIPV modules

Preventive maintenance is important to avoid potential failures and malfunctions of the BIPV modules and, thus, systems. During maintenance, the following recommendations should be considered:

Accessibility: The BIPV modules should be accessible for maintenance.

Soiling: Soiling and debris depend on local sources of dirt and rain frequency. Care must be taken with the cleaning of PV modules to avoid damage and follow the manufacturer's recommendations (generally, the use of plain demineralised water and mild detergent is recommended while the use of high-pressure water, brushes, or any types of solvents, abrasives, or harsh detergents should be avoided).

Snow: While snow can lead to a reduction in system performance and complicates the provision of O&M services, it is generally not recommended to be removed from the modules due to the damaging potential.

Anchors and fastening: BIPV modules connections to the building structure should be periodically inspected.

Temperature: BIPV modules temperature should be periodically evaluated. Thermographic tools are useful for a global temperature monitoring of the system.

Electrical performance and fault detection: Several actions can be taken to facilitate a complete assessment of the PV modules' performance. The use of current-voltage curve measurements and visual inspection (e.g., for glass breakage, burn marks, delamination, browning, or damaged backsheets) can be performed. On-site inspection methods with portable test equipment are thoroughly described in [4].

6.5.2 Wiring

O&M measures and costs depend on the wire management system employed. Preventive maintenance entails selecting safe and robust wires. Corrective maintenance entails finding and repairing ground faults, such as wire scraping against a module frame or arc faults caused by a broken wire or loose connection. Wire management can be classified into three categories: open air, direct bury, and conduit/cable tray.

Electrical connections should be periodically checked by visual inspection and thermographic inspection or impedance analysis of the circuit. Ground connections, fuses, breakers, and arc detectors should be periodically tested.

6.5.3 Inverter

The DC/AC power conversion equipment (inverter) has a great impact on the BIPV system reliability. Sticking to the manufacturer's O&M instructions is the preferred preventive option. That includes, e.g., tightening fasteners on terminal blocks for current-carrying conductors and conducting thermal imaging of electrical connections and components. Note that the O&M procedures depend on the inverter topology, which ranges from microinverters on individual modules, string inverters, DC-optimizers, or larger central inverters.

6.5.4 O&M considerations for energy storage systems

Most used energy storage systems (ESS) are electrochemical batteries. The end user must follow the instructions supplied by the ESS manufacturer. The following list provides recommended O&M actions:

Visual Inspection: Periodic checks should be made for dents, swelling, or discoloration in the ESS. The positive and negative terminals of the power connector should be ensured to be free from oxidation.

Voltage Checking: The ESS voltage should be regularly monitored, ensuring it remains within the rated range.

Idle or Stored Batteries: In the case of batteries being idle or stored, the state of charge (SoC) should be maintained within a specified range through periodic charging.

Periodic Full Charge: Consideration can be given to performing periodic full charging using the energy management system to update the state of charge (SoC) and prevent measurement errors.

Component Check: Inspections should be conducted on power connectors, cables, grounding points, and screws to detect any signs of looseness, damage, or corrosion.

Battery Placement: Batteries should be kept away from direct sunlight and rainfall and in accordance with the manufacturer's ambient specifications (e.g., temperature, relative humidity and ventilation rates).

Safety Environment: The installation environment should be kept free from safety hazards to the ESS, such as dust, water, rodents, insects, etc.

Furthermore, batteries should be positioned away from frequented spaces within the building to ensure quick and safe egress in the event of a fire or emergency. The selection of an appropriate location for batteries is vital to prevent health and safety risks.

References

[1] SolarPower Europe, *Best Practice Guidelines Maintenance & Operation – Version 5.0*, 2021 [Online]. Available: www.solarpowereurope.org.

[2] N. Renewable Energy Laboratory, S. National Laboratory, S. Alliance, and S. National Laboratory Multiyear Partnership, *Best Practices for Operation and Maintenance of Photovoltaic and Energy Storage Systems*, 3rd edition, 2018 [Online]. Available: www.nrel.gov/publications.

[3] J. Ulrike, B. Herteleer, C. Tjengdravira, et al., "Guidelines for operation and maintenance of photovoltaic power plants in different climates 2022 PVPS task 13 reliability and performance of photovoltaic systems. Report IEA-PVPS T13-25:2022," 2022 [Online]. Available: https://iea-pvps.org/research-tasks/performance-operation-and-reliability-of-photovoltaic-systems/.

[4] K. Pickerel, "Just how concerned should the solar industry be about battery fires?," *Solar Power World*, pp. 42–45, 2020 [Online]. Available: www.solarpowerworldonline.com/2020/02/just-how-concerned-should-the-solar-industry-be-about-battery-fires/.

ANNEX I

Symbols, acronyms, and abbreviations

The guidebook follows the overall rule that symbols representing variables are in italics, and symbols representing units, or labels, are in roman fonts. Acronyms and abbreviations are in roman too. The following alphabetical list contains symbols, acronyms and abbreviations used in the text:

α_{sol}	solar absorptance (%)
β	voltage temperature coefficient (variation of PV power with temperature difference) (%/K)
γ	power temperature coefficient (variation of PV power with temperature difference) (%/K)
σ	Stefan-Boltzmann constant: $5.6703 \cdot 10{-}^{8}$ W/($m^2 \cdot K^4$)
$\tau(\lambda)$	spectral transmittance λ, wavelength
τ_{sol}	solar transmittance (%)
τ_{vis}	visible transmittance (%)
ΔT	temperature difference (K, °C)
a, *b*	empirical coefficients of Sandia temperature model [*a* dimensionless, *b* $(m/s)^{-1}$]
AC	alternating current
AM	air mass
A_m	total area of a PV module (BIPV module), defined by its outermost edges
ARC	anti-reflective coating
a-Si	amorphous silicon
BAPV	building-added photovoltaic(s) (or building-attached photovoltaic(s))
BIPV	building-integrated photovoltaic(s)
BPS	building performance simulation
CdTe	cadmium telluride
CEI	International Commission on Illumination
CI(G)S	copper indium (gallium) selenide
CRI	color rendering index (%)
DC	direct current
DSF	double-skin façade
DSM	demand-side management
E_{PV}	photovoltaic produced energy (kWh)
EVA	ethylene-vinyl acetate
G	total solar irradiance (W/m^2)
GHG	green-house gases
GL	glass laminate (laminated glass)
$\mathrm{G_{POA}}$	solar irradiance in the plane of array (W/m^2)
G_{ref}	reference solar irradiance, typically 1 kW/m^2 (STC)
$\mathrm{H_{POA}}$	solar irradiation in the plane of array (kWh/m^2)
HVAC	heating, ventilation and air-conditioning
IEA	international energy agency

IEC	International Electrotechnical Commission
IGU	insulated glass unit
ISO	international organisation for standardisation
k	Ross model coefficient (m^2K/W)
KPI	key performance indicator
LCA	life cycle analysis
LCOE	levelised cost of energy
L_i	electric power loss due to cause "i"
LID	light-induced degradation
L_{temp}	electric power loss due to temperature increase
mc-Si	multicrystalline silicon
MJSC	multi-junction solar cell
MPP	maximum power point (refers to electrical *I-V* characteristics of PV devices)
m-Si	monocrystalline silicon (single-crystal silicon)
NPV	net present value
NMOT	nominal module operating temperature (°C)
NOCT	nominal operating cell temperature (°C)
NZEB	net-zero energy building
OC	open circuit
O&M	operation and maintenance
P_0	nominal power of a PV or BIPV system (summation of maximum output power at STC of all the installed modules)
PID	potential-induced degradation
P_m	power of a PV or BIPV module (W)
P_{max}	maximum power of a PV or BIPV module at STC (W)
POA	plane of array (plane of the modules)
POE	polyolefin elastomer
PR	performance ratio
PV	photovoltaic, photovoltaics
PVB	polyvinyl butyral
PVDF	polyvinylidene difluoride
PVDG	photovoltaic double-glazing
PVGL	PV glass-glass laminate (photovoltaic laminated glass)
SC	short circuit
SCI	self-consumption index
SHGC	solar heat gain coefficient (or g value) (%)
SHJ	silicon heterojunction
Si	silicon
SSG	structural sealant glazing
SSI	self-sufficiency index
STC	standard test conditions (irradiance 1 kWh/m^2, ambient temperature 25 °C, solar spectrum AM1.5G)
STPV	semi-transparent PV
T	temperature (K, °C)
T_a	ambient temperature (K, °C)
T_{back}	PV module back surface temperature (K, °C)
T_c	PV module cell temperature (K, °C)
TCO	transparent conductive oxide
T_m	PV module temperature (K, °C)
U	thermal transmittance ($W/(m^2K)$)
w_s	wind speed (m/s)

$Y_{r,}$	reference yield (equivalent number of hours at 1 kWh/m^2) (units: h or $(kWh/m^2)/(kW/m^2)$)
η	PV efficiency
τ_{sol}	solar transmittance
τ_{vis}	visible transmittance
$\rho(\lambda)$	spectral reflectance; λ, wavelength

ANNEX II

List of referenced standards

The list contains the name of the last published edition of each standard or technical specification. Some of them may be under review at the date of the publication of the guidebook. The important information is the name of each standard, the date will change as new editions are published.

- IEC 63092–1:2020 Photovoltaics in buildings – Part 1: Requirements for building-integrated photovoltaic modules
- IEC 63092–2:2020 Photovoltaics in buildings – Part 2: Requirements for building-integrated photovoltaic systems
- IEC 60904–1:2020 Photovoltaic devices – Part 1: Measurement of photovoltaic current-voltage characteristics
- IEC 61215–2:2021 Terrestrial photovoltaic (PV) modules – Design qualification and type approval – Part 2: Test procedures
- IEC 61730–1:2023 Photovoltaic (PV) module safety qualification – Part 1: Requirements for construction
- IEC 61730–2:2023 RLV Photovoltaic (PV) module safety qualification – Part 2: Requirements for testing
- IEC TS 62804–1:2015 Photovoltaic (PV) modules – Test methods for the detection of potential-induced degradation – Part 1: Crystalline silicon
- IEC 60364 (series) Low-voltage electrical installations
- IEC 61724–1:2021 RLV Photovoltaic system performance – Part 1: Monitoring
- ISO 10291: 1994 Glass in building. Determination of steady-state U values (thermal transmittance) of multiple glazing. Guarded hot plate method
- ISO 10292:1994 Glass in building. Calculation of steady-state U values (thermal transmittance) of multiple glazing
- ISO 10293:1997 Determination of steady-state U values (thermal transmittance) of multiple glazing. Heat flow meter method
- ISO 9050:2003 Glass in building – Determination of light transmittance, solar direct transmittance, total solar energy transmittance, ultraviolet transmittance and related glazing factors
- ISO 15099:2003 Thermal performance of windows, doors and shading devices – Detailed calculations
- ISO 19467:2017 Thermal performance of windows and doors – Determination of solar heat gain coefficient using solar simulator
- ISO 8995–1:2002 Lighting of work places. Part 1: Indoor
- ISO 12543 1–6 series Glass in building – Laminated glass and laminated safety glass
- ISO 29584: 2015 Glass in building. Pendulum impact testing and classification of safety glass
- ISO/TS 18178: 2018 Glass in building. Laminated solar photovoltaic glass for use in buildings
- ISO 11925–2:2020 Reaction to fire tests. Ignitability of products subjected to direct impingement of flame. Part 2: Single-flame source test
- ISO 834 (series) Fire-resistance tests. Elements of building construction
- EN 410:2011 Glass in building – Determination of luminous and solar characteristics of glazing
- EN 50583-1:2016 Photovoltaics in buildings - Part 1: BIPV modules
- EN 50583-2:2016 Photovoltaics in buildings - Part 2: BIPV systems

Index

acoustic insulation 11, 33
adaptability 51
aesthetics 16–17, 29, 43, 47
airflow 36, 48
airtightness 12, 42
amorphous silicon (a-Si) 4–5, 11, 25
annual energy yield 5, 43–45, 47
annual solar access study 44–45
anti-reflective coating (ARC) 14, 29, 30–31
architectural compatibility 17
architectural integration 26, 34
architectural objectives 17
array 47, 49–50
atrium 7
awareness 17

backsheet 27–30, 31–32
balance of system 49, 52
balustrade 34, 39
battery 49–50,172
bifacial PV cells 25, 30, 36, 39
BIPV-as-a-service 54
BIPV design workflow 42–43
BIPV energy generation target 45, 47
BIPV module efficiency 3–4, 30
BIPV system efficiency 5
building-attached photovoltaics (BAPV) 8, 28
building code 13, 28, 44
business model 51, 53–54
bylaws 44

cadmium telluride (CdTe) 4, 25, 51–52
carbon-neutral(ity) 46
Carport canopy 160–166
cell string 25–26, 30
circular business model 51
circularity 50–51
cladding 34–36
colour(ed) 4, 9–10, 17, 29–30, 32
colour rendering 9–10
comfort 9, 37, 47
construction documents 43, 48
construction permit 50
construction stages 42
continuous roof 32–33
copper indium gallium selenide (CIGS) 4–5, 25, 37, 39, 51–52
corrosion 14, 26, 30
crystalline silicon (c-Si) 4, 24–25, 30
current 6, 14, 16, 24–25
curtain wall 12, 36, 47–48, 136–150
customisation 26–27

daylight(ing) 9, 20, 31, 37, 47
decarbonisation 1, 43
decision-making process 42
degradation 13–14
demand-side management (DSM) 7, 46
design 42, 56
design for disassembly 51
design stage 43
discontinuous roof 10, 32, 58–87
documentation 51, 169
double-skin façade 34, 36, 128–131
durability 8, 12–13, 51

economic viability 52
efficiency 3–5, 13, 25, 30–31, 46
electrical design 5–6, 58, 88, 106, 136, 160
electricity generation 3, 42–45, 47–48
electrification 43
encapsulation 26, 30
energy storage 6–7, 172
energy yield 5, 43–45, 47, 171
enhanced aesthetics 43
environmental footprint 51

ethylene vinyl acetate (EVA) 15, 30
Evan's model 47

façade 5, 17–20, 33–36, 44, 106–159
Faiman's model 8
fastening 51, 171
fault detection 171
feed-in tariff 53
fire safety 14–15, 36, 50, 58, 88, 106, 136, 160
flat roof 15, 33
flexible BIPV 29–30, 33, 37
frame (or framing) 4–5, 12, 16, 26, 28
frequency 49
functional unit 24

glass-glass module (or laminate) 25, 28, 30, 36, 51
greenhouse gas (GHG) emissions 46, 52
grid-connected BIPV system 48–50, 51, 52, 53
ground(ing) 14, 15, 49
ground-mounted 5–6, 47

hail 13
heating, ventilation and air conditioning (HVAC) 47–48
heat transfer 8–9
heritage 43–44
high efficiency 6, 27, 31
hotspot 5–6, 14–15, 25, 48–49
hygro-thermal 47, 58, 88, 106, 136, 160

impact resistance 12
incentive 43, 53–54
International Energy Agency (IEA) 3, 20
inverter 6, 25, 41, 48–50, 169, 172
investment 52–53

laminated glass 11, 13, 26
lamination 13, 26, 30
levelised cost of energy (LCOE) 52
life cycle cost analysis 52
life cycle metrics 52
light-induced degradation (LID) 14
low-emissivity 8, 25

maintenance 51–52, 53, 54, 171
manufacturing 25–29, 51
mass customisation 27
mass production 26
mechanical stability 13
microinverter 6, 49
module 6, 8, 15–16, 25–31, 48, 171–172
module manufacturing 25–28
module temperature 8, 32
monocrystalline silicon 4, 24
mounting system 48
multicrystalline silicon 4, 24
multijunction solar cells (MJSC) 25

net-metering 42, 53
net-zero energy 43, 46
nominal operating cell temperature (NOCT) 8
nominal power 28

one-diode model 48
operation and maintenance (O&M) 2, 54, 167–172
optimizer 49, 50
orientation 17, 43

parapet 39
partial shading 5–6
pattern 7, 17, 29
payback period (PP) 52
perception 17
performance indicator 51
performance ratio (PR) 5, 171
placement and layout 17
polyvinyl butyral (PVB) 30, 34
power conversion equipment 49
power loss 3–5, 14
power purchase agreement (PPA) 53–54

rain(fall) 28, 48
rainscreen 35, 47, 106–135
reflectivity 29
regulatory compliance 43, 169
renewable energy generation 37
retrofit 45, 50
reversibility 51
roof(ing) 10–11, 32–33, 47, 58–87
Ross model 8

safety 12–14, 28, 169
Sandia model 8, 48
seismic load 12
self-consumption 6–7, 34, 46, 53
self-sufficiency 6–7, 46
semi-transparent 9–10, 25, 30
shading 5–6, 36–39, 40, 48–49, 151–159, 169
silicon heterojunction (SHJ) 4, 25
site 43–44

skylight 33, 47, 88–105
snow 48, 171
soiling 3, 5, 48, 171
solar access study 45–46
solar availability 44–45
solar cell 4, 24–25, 29–35
solar heat gain coefficient (SHGC) 9
solar transmittance (Tsol) 9, 29, 30
soldering 26, 51
string 6, 25–26, 49
structural design 58, 88, 106, 136, 160
sustainability 17, 43, 50, 58, 88, 106, 136, 160
system configuration 8, 48, 50

tandem cell 25
temperature 5, 7–8, 12
thermal performance 7, 47
thermal stress resistance 12
thermomechanical degradation 13
thin-film PV cells 25, 51, 52
tilt angle 5, 43–44

transmittance 7, 9, 29
transparency 17, 20, 29, 32
transparent conductive
  oxide (TCO) 25

utility 44, 46, 50
U value 7–8

ventilated façade 7–8, 35
ventilation 3, 8, 42
visual comfort 9, 37
voltage 6, 14–15, 49

wafer 24–25
walkable 33–34
water control 3, 10
weatherproofing 3, 10
wind load resistance 12
wiring 28, 49, 169, 171–172

zoning 44

For Product Safety Concerns and Information please contact our EU
representative GPSR@taylorandfrancis.com
Taylor & Francis Verlag GmbH, Kaufingerstraße 24, 80331 München, Germany

www.ingramcontent.com/pod-product-compliance
Ingram Content Group UK Ltd.
Pitfield, Milton Keynes, MK11 3LW, UK
UKHW050757180926
475701UK00007B/1100
* 9 7 8 1 0 3 2 5 1 7 0 5 6 *